A World List of Mammalian Species

Third edition

A World List of Mammalian Species

Third edition

G.B. Corbet & J.E. Hill

Illustrated by Ray Burrows

Natural History Museum Publications

OXFORD UNIVERSITY PRESS

1991

A World List of Mammalian Species

Third edition 1991

Published by
Natural History Museum Publications
Cromwell Road, London SW7 5BD

&

Oxford University Press, Walton Street, Oxford OX2 6DP
Oxford New York Toronto
Delhi Bombay Calcutta Madras Karachi
Petaling Java Singapore Hong Kong Tokyo
Nairobi Dar es Salaam Cape Town
Melbourne Auckland

and associated companies in
Berlin Ibadan

Oxford is a trade mark of Oxford University Press

Published in the United States
by Oxford University Press, New York

British Library Cataloguing in Publication data
Corbet, G.B. (Gordon Barclay) 1933-
World list of mammalian species.-3rd ed.
1. Mammals
I. Title J.E. Hill, John Edwards 1928-
599 00216

ISBN 0-19-854017-5

Library of Congress Cataloging-in-Publication Data
Corbet, G.B. (Gordon Barclay)
A world list of mammalian species/G.B. Corbet and J.E. Hill. – 3rd ed. p. cm.
Includes bibliographical references (p.) and index.
1. Mammals-Nomenclature. 2. Mammals-Classification.
I. Hill, John Edwards. II. Title.
QL708.CS7 1991 559'. 0012-dc20 90-7790

Typeset in Times by S.B. Data Graphics Ltd, Colchester, England.
Printed by St Edmundsbury Press, Suffolk, England.

Contents

Introduction

After birds, mammals are the group of animals whose diversity has been most intensively studied. But the attention of zoologists, whether concentrating on structure, ecology, behaviour or physiology, has been very unevenly distributed amongst the four thousand or so species that make up this assemblage. Some, like the chimpanzee, have had volumes devoted to every aspect of their structure and lives; at the other extreme species have been described on the basis of such flimsy evidence as to leave their very existence in doubt.

What is a mammalian species?

In devising a classification of mammals, or any other group of animals, the basic task is to detect, amongst the bewildering variety of individual animals, those genetically independent groups that we call species and that the animals themselves would recognize as constituting 'us' as distinct from 'them'. This concept of the species is not absolutely rigid and clear-cut but it is based upon the fact that individuals within a geographically coherent group show a degree of resemblance to each other that arises from a common ancestry and allows them to interbreed freely, thereby disseminating their characteristics within the group. The difference between members of such a species and members of the most closely related species occupying the same area may be conspicuous or obscure – it is the absence of intergradation rather than the magnitude of the difference that enables species to be defined. Given adequate information there is rarely difficulty in defining the species at any one locality, but the concept of the species becomes less precise when we consider geographically separated populations and have to speculate as to whether any differences between them are sufficiently slight that they would interbreed freely if reunited.

In some groups of mammals, for example the ungulates and carnivores, it is unlikely that many completely new species await discovery and our delimitation of known species is unlikely to change much. However some uncertainties are almost bound to remain since species are continually evolving and it is almost inevitable that some pairs of populations have diverged to just such an extent that it is rather arbitrary whether they are deemed to have crossed the species threshold and been irretrievably set upon separate evolutionary pathways. Amongst the antelopes for example the four principal forms of oryx in Africa and Arabia, all geographically isolated from each other, have at one extreme been considered races of a single species, *Oryx gazella*, but are more often treated as three species, as here (p.132), or even four.

Amongst the small mammals the situation is very different. Several completely new species, mainly bats and rodents, are discovered each year. Known species are sometimes found, on closer inspection, to comprise two or more separate but very similar species – 'sibling species' – whose genetical independence may sometimes be confirmed by the degree of difference in their chromosome complements or by biochemical studies. However, because these techniques cannot usually be applied to existing collections in museums, it frequently happens that the samples available for study are so small or so unrepresentative of the species as a whole that the results are equivocal and it may not be practicable to use them in compiling a list such as this. On the other hand any such proliferation of known species tends to be counter-balanced by the discovery that other 'nominal species', i.e. forms that have been described and named as species on the basis of inadequate samples, are in fact only variants of other, better known, species.

The scope of the list

This book is an attempt to present a comprehensive list of all living species of mammals as far as current knowledge allows. Recently extinct species are also included, marked †, if their external appearance is known from preserved specimens, illustrations or descriptions, e.g. the quagga, an extinct zebra from southern Africa, but not if they are known only from skeletal remains. (An exception is made in the case of the beaked whale *Mesoplodon pacificus*, at present known only from two skulls but quite likely to be surviving.) A number of 'nominal species' are also omitted if it seems unlikely that they represent independent species even if they cannot be confidently allocated to a particular known species (usually because they are based on inadequate descriptions and the original specimens are lost or poorly preserved). The best place to find more information on any species is generally the most recent regional work, for example those listed on pages 213 to 215. For very few groups of mammals are there comprehensive descriptive works that ignore national frontiers.

Higher classification and sequence

It is not intended that this book should be considered an original source for the higher classification of mammals, i.e. classification above the level of genus. The classification used is based, like those in most recent compilations, on that of Simpson published in 1945 (see p.213), adapted to take account of more recent work. The result is a compromise between keeping a well-known classification in spite of the fact that subsequent work has shown some aspects of it to provide a poor indication of relationships, and recent alternatives that have not yet had time to be adequately tested. Ranks other than order, family and genus have been avoided except in the Chiroptera and in a few other cases where they are particularly clear-cut and usefully divide large groups, e.g. in the rodent family Muridae where subfamilies have been used.

We have likewise followed the *sequence* used by Simpson unless the requirements of a more recent classification have directed otherwise. At the level of the orders we have adopted the principal conclusions of the phylogenetic classification of McKenna (see p.213) but have minimized changes of sequence and rank. The main effect of this is to separate the order Edentata from the rest of the placental mammals and within the remainder to recognize the Macroscelidea and the Lagomorpha as comprising a pair of distinctive groups. There are severe limits on the extent to which a linear sequence can reflect the tree of relationships uniting a multitude of groups. We have therefore adopted a simple alphabetical sequence of species within each genus. This has the disadvantage of separating many closely related species but any simple alternative could equally mislead in implying, by juxtaposition, affinities that were not intended. The disadvantages of a non-alphabetical arrangement of genera and higher categories are alleviated by the provision of an index.

Number of species

In quoting the number of species in an order or family the form '*c*. 23 species' is generally used, meaning that the number of species *listed* is exactly 23 but that there is sufficient uncertainty about the status or content of some of them to make it likely that this number will be revised as the species become better known, or will remain unstable because of species that are marginally distinct. Absence of the '*c*.' implies that the species are clearly defined and the number is unlikely to change.

Nomenclature

No authors nor dates are given with the scientific names – these can be found in the checklist by Honacki *et al.* (see p.213) and in many of the other references quoted. Except when dealing with problems of nomenclature, or with recently described species, it is usually quite unnecessary to quote the author or date of the name of a mammal. Again with the exception of recently described species, the original citation is rarely the best place to go to find a useful description of the species.

The name of the genus is followed, in parenthesis, by any synonyms that are frequently used in recent literature. Zoologists should note that these 'synonyms' are simply alternative generic names that are commonly used for all or any of the species listed in the genus – it is not necessarily implied that they are formal junior synonyms of the generic names used. The same applies to synonyms of species names – these most often represent forms that have recently been considered separate species but are here included in the other, earlier-named, species. For example on p.172 '*Gerbillus* (*Dipodillus, Monodia*)' indicates that some of the species listed under *Gerbillus* have at times been placed in separate genera *Dipodillus* or *Monodia* (which are not listed here as accepted genera). On page 174 '*Gerbillurus* (*Gerbillus*)' indicates that the species here listed under the genus *Gerbillurus* have at times been included in the preceding genus *Gerbillus*.

Vernacular names, in English only, are given if they are well established or if they are used in a major work on the region or group concerned. We have invented a few names, especially in cases where an obvious name suggested itself by analogy with a related species that already had an accepted English name. We have also enlarged some well established names to make them unique, e.g. Eurasian red squirrel, without implying that the expanded version need always be used. However, many vernacular names, especially of bats and rodents, are only unambiguous when used in a local context.

Geographical range

The aim has been to be as precise as possible within the severe limitations of space, helped in many cases by an indication of the vegetation zone or zones concerned. Small islands are generally ignored unless they comprise a particularly significant part of the range. The form 'Uganda, etc.' is used to mean that the range is centred on Uganda but extends slightly into adjacent countries. In south-eastern Asia the expression 'Burma – Java' implies presence in Malaya and Sumatra but not Borneo and Philippines unless these are specifically mentioned. Parentheses ( ) are used to indicate a minor component of the range, especially in the case of genera or higher groups. For example in the case of the opossums of the family Didelphidae 'S, C America, (N America)' reflects the fact that there are many species in south and central America but only one in north America. Brackets [ ] indicate populations introduced by man, but are generally used only in the case of very extensive and well established populations remote from the original range, e.g. coypus introduced from South America into Eurasia and European rabbits in Australia. Introductions are ignored in giving the range of higher groups. In the case of a species in which a major diminution of range over the last hundred years or so has been well documented current range is given but areas formerly occupied are indicated where practicable. No attempt has however been made to indicate earlier ranges known only from fragmentary information or subfossil remains. Extinct species are marked † against the scientific name as well as 'extinct' following the range.

Habitat and ecology

This information is not intended to be comprehensive but is used especially to define the range more precisely. Where habitat is constant for a genus or higher group it is not repeated under each species. Expressions such as 'forest' or 'steppe' are intended to define the gross biotic zone that the species occupies rather than the habitat selected by individuals. Generalized descriptive terms for structural vegetation types are used in preference to those with local or floristic connotations, e.g. 'scrub' rather than 'chaparral'. The principal terms used for vegetation types can be defined as follows:

Forest – dominated by trees. Qualified as closed, open, rain, dry, deciduous, evergreen, coniferous.

Woodland – sometimes used here for open forest but generally avoided because of ambiguity. It can also mean secondary forest or small scattered segments of forest.

Scrub – closed shrub vegetation.

Savanna – closed grassland (long or short) with scattered trees.

Steppe – open grassland, herb or dwarf shrub (i.e. bare ground between plants).

Desert – bare ground predominating.

Terms such as 'arboreal' and 'subterranean' are self-explanatory – the latter implies that the animals find most of their food underground and is not used merely to indicate the construction of burrows for shelter. 'Terrestrial' is used to mean living mainly on the ground surface in contrast to 'arboreal' and 'subterranean' as well as to 'aquatic'.

Feeding habits are generally given for higher categories rather than for individual species and then only if they are reasonably constant for the group.

Sources and references

The reference numbers refer to the bibliographies on pages 216 to 231. Since the major sources of information are generally works covering all groups of mammals in a limited geographical region, these are listed in a separate bibliography (p.213) and are not referred to in the text of the list. It can generally be assumed that the principal authority for the recognition of a particular species is the most recent publication in the appropriate geographical bibliography, unless there is in the text of the list a reference to a work dealing with the particular species or higher taxonomic group. Particular attention has been paid to documenting the source for species that are not listed by Honacki *et al.* (p.213), nor in the latest regional works, and to providing references to the original descriptions of genera and species that have been newly described and named during the last ten years (as distinct from those that have simply been reinstated as species).

These references are intended primarily to indicate sources where further detail and justification can be found for the acceptance of the species included in the list, e.g. in deciding such issues as whether three named forms should be considered as belonging to one species or to three separate species. They have not necessarily been followed with regard to other aspects such as nomenclature and higher classification. The bibliographies also contain supplementary sources, not referred to in the text, which provide useful additional information but have not been used as authorities for the recognition of species.

Endangered species

For those species that are listed in the *IUCN red book of threatened animals* (see bibliography, p. 213) the entry is concluded by a letter indicating the category used in the IUCN list. These are as follows:

E   Endangered
V   Vulnerable
R   Rare
I   Indeterminate (known to be E, V or R)
K   Insufficiently known (but suspected of qualifying for E, V or R)
T   Threatened (comprises subspecies or populations with different status)

Enclosure of a letter in parentheses ( ) indicates that it refers to only part of a species, e.g. a particular subspecies or a particular part of the range.

Domesticated mammals

The naming of domesticated animals and their wild ancestors is confused. On the one hand, both wild and domesticated forms have generally been given separate formal scientific names (e.g. *Canis lupus* for the wolf and *Canis familiaris* for the dog). But it has also been argued that because wild and domestic forms are normally interfertile they should be considered as belonging to the same species. In this case the valid name for the species may be that given originally to the wild form or to the domestic form, depending upon which was the first to be proposed. This leads to many confusing anomalies that conflict with general usage and various proposals have been made to resolve the problem. None has found general acceptance.

The solution followed here is to consider that since clearly distinguishable domesticated forms do not in fact interbreed with the wild species to the extent of losing their separate identity (even although they are potentially capable of doing so), they should not be considered a part of the ancestral wild species and consequently the names applied to domesticated forms should not be used as the names of the wild species. Only the wild species are listed here, but any domesticated derivatives are noted after the range. Names based on domesticated forms are given as synonyms in the first column only when they have frequently been used to include the wild species. An exception is made only in the case of the dromedary which is included in the list since no distinct ancestral species is known.

For a more detailed account of the nomenclature of domesticated animals see the appendix in *Evolution of domesticated animals* edited by I.L. Mason (Longman, 1984).

Extinct species

The 27 extinct species listed below are also included in the main list marked †. Although extinction is all-or-nothing from the point of view of the species concerned, our knowledge of which species are extinct is much less precise. If an inconspicuous species has only ever been found on a few occasions how long should one wait after the last report before considering it likely to be extinct? Hopefully some of these may yet be discovered alive; but it is also possible that many more should already be on the list if we but knew.

Note that these are all species that have been known 'in the flesh'—neither this nor the main list includes species known only from fossil or subfossil skeletal remains.

Thylacinus cynocephalus   Thylacine
Perameles eremiana   Desert bandicoot
Chaeropus ecaudatus   Pig-footed bandicoot
Macrotis lecura   Lesser rabbit-bandicoot
Potorous platyops   Broad-faced potoroo
Caloprymnus campestris   Desert rat-kangaroo
Lagorchestes leporides   Eastern hare-wallaby
Macropus greyi   Toolache wallaby
Pteropus subniger   Lesser Mascarene flying fox
Pteropus tokudae   Guam flying fox
Phyllonycteris major   Puerto Rican flower bat
Mystacina robusta   New Zealand greater short-tailed bat
Dusicyon australis   Falkland Island wolf
Monachus tropicalis   Caribbean monk seal
Hydrodamalis gigas   Steller's sea cow
Equus quagga   Quagga
Cervus schomburgki   Schomburgk's deer
Hippotragus leucophaeus   Blue buck
Gazella rufina   Red gazelle
Oryzomys victus   St Vincent rice rat
Nesoryzomys darwini ⎫
Nesoryzomys indeffesus ⎬ Galapagos rats
Nesoryzomys swarthi ⎭
Pitymys bavaricus   Bavarian pine vole
Rattus macleari ⎫
Rattus nativitatis ⎬ Christmas Island rats
Crateromys paulus Ilin bushy-tailed cloud rat

Distribution of mammalian orders by zoogeographical region

This table shows the approximate number of species of each order of mammals in each of the major zoogeographical regions. It is limited to species that occur primarily or extensively in the region concerned. The numbers are approximate:

	Palaearctic	Nearctic	Neotropical	Afrotropical	Indomalayan	Australasian
Monotremata						3
Marsupialia		1	82		2	195
Xenarthra		1	29			
Insectivora	66	48	10	155	52	
Scandentia					16	
Dermoptera					2	
Chiroptera	61	50	232	175	251	127
Primates	2		64	81	52	
Carnivora	38	30	47	72	65	
Proboscidea				1	1	
Perissodactyla	3		3	5	4	
Hyracoidea	1			8		
Tubulidentata				1		
Artiodactyla	39	11	15	76	34	
Pholidota				4	3	
Rodentia	239	200	450	265	304	115
Lagomorpha	25	15	5	11	6	
Macroscelidea	1			14		
Total	475	356	937	868	792	440

species confined to transitional areas have been ignored; species extending marginally into a region from their principal region have been ignored; those that occur extensively in more than one region are counted for each; oceanic islands have been ignored.

The regions are defined roughly as follows:
Palaearctic: Europe, Asia north of the Himalayas and central China, Africa north of the Sahara.

Nearctic: N America north of central Mexico.
Neotropical: Central and South America.
Afrotropical: Africa south of the Sahara, including Madagascar.
Indomalayan: SE Asia from the Himalayas and central China to the Philippines
              and Sulawesi.
Australasia: Australia, New Guinea and adjacent islands.

Authorship

In compiling the list the bats (order Chiroptera) have been dealt with by J.E. Hill
and the other orders by G.B. Corbet.

Note on the third edition

In bringing the list up to date we have concentrated on adding or deleting species
according to recently published research. These changes result from the
publication of taxonomic revisions and new or revised regional compilations, and
from the description of new species, resulting in a net increase of 93 species.
Changes at higher levels of classification have been kept to a minimum. Many
aspects of the higher classification used here (i.e. at the level of genus, family and
order as distinct from the delimitation of species) are becoming increasingly out of
date. However, much of the recent research in this field, using for example
biochemical and numerical methods, has not been sufficiently comprehensive to
permit the allocation of all known members of the group to the new classification,
or has resulted in classifications that are only marginally more likely to reflect
phylogenetic relationships than the existing one, or indeed than many other
alternatives. There is therefore a need for a period of consolidation and testing
before such revisions can be usefully incorporated into a list such as this.
Some changes in the insectivores (Insectivora), tree shrews (Scandentia), rodents
(Rodentia) and bats (Chiroptera) of the Indomalayan region result from a more
detailed review of the mammals of that region that we are currently preparing.

The World List of Mammalian Species

ORDER MONOTREMATA

Monotremes; 3 species; Australia, New Guinea; terrestrial and aquatic predators, mainly on invertebrates.

Family Tachyglossidae

Spiny anteaters (echidnas); 2 species; Australia, New Guinea; terrestrial insectivores.

Short-nosed echidna (Tachyglossus)

Tachyglossus

T. aculeatus	Short-nosed echidna	Australia, Tasmania, SE New Guinea; steppe, forest

Zaglossus

Z. bruijnii (*bartoni*)	Long-nosed echidna	New Guinea; forest; V

Family Ornithorhynchidae

One species; a freshwater predator.

Ornithorhynchus

O. anatinus	Platypus	E Australia, Tasmania

ORDER MARSUPIALIA

Marsupials; *c.* 282 species; Australia, New Guinea, (Sulawesi), S America, C America, (N America); all terrestrial habitats; herbivores and predators; ref. 1.1.

Platypus (Oraithorhyachus)

Family Didelphidae

American opossums; *c.* 75 species; S, C America, (N America); mainly forest; predators and omnivores.

Marmosa; mouse-opossums; S, C America; forest, grassland.

M. andersoni		Peru
M. canescens	Greyish mouse-opossum	S Mexico
M. cracens		Venezuela
M. fuscata (*carri*)		Colombia, W Venezuela, Trinidad
M. handleyi		Colombia; ref. 1.12
M. impavida	Pale mouse-opossum	E Panama – Peru, W Brazil
M. incana		E Brazil
M. invicta	Panama mouse-opossum	Panama

Mouse opossum (Marmosa murina)

M. janetta		W Paraguay, Bolivia; ref. 1.17
M. leucastra		N Peru
M. mexicana	Mexican mouse-opossum	Mexico – Panama
M. murina		Tropical S America, Tobago
M. noctivaga		Tropical S America
M. ocellata		Bolivia
M. parvidens		Brazil, Peru – Surinam, Colombia
M. quichua		E Peru
M. robinsoni (*mitis*)		Belize – NW South America, Trinidad, Tobago, Grenada
M. rubra		E Ecuador, S Peru
M. scapulata		SE Brazil
M. tyleriana		Venezuela
M. xerophila		Venezuela, Colombia
M. yungasensis		W Bolivia – Ecuador

Micoureus; (*Marmosa*); ref. 1.17.

M. alstoni (*cinerea*)	Colombia – Honduras
M. cinerea	N Argentina – Colombia
M. constantiae	C Brazil – N Argentina
M. domina	Amazonian Brazil
M. germana	E Ecuador, Peru
M. mapirensis	Bolivia, Peru
M. phaea	W Colombia, W Ecuador
M. rapposa	SE Peru, NE Bolivia
M. regina	Colombia

Thylamys; (*Marmosa*); ref. 1.27

T. elegans	NW Argentina, Bolivia, Peru, Chile
T. grisea	Paraguay
T. karimii	C Brazil
T. lepida	Amazon Basin, Surinam
T. pusilla	Argentina, Bolivia, Paraguay
T. tatei	Peru
T. velutina	SE Brazil

Gracilinanus; (*Marmosa*); ref. 1.28.

G. aceramarcae	Bolivia
G. agilis	Central S America
G. dryas	W Venezuela, E Colombia
G. emiliae	E Brazil

G. marica		N Venezuela; montane
G. microtarsus		SE Brazil, NE Argentina

Monodelphis; short-tailed opossums; S, C America; forest, savanna.

M. adusta	Cloudy short-tailed opossum	Panama – Peru
M. americana	Three-striped short-tailed opossum	Guianas, Brazil, NE Argentina
M. brevicaudata (*touan*)	Seba's short-tailed opossum	Colombia – Surinam – N Argentina; forest
M. dimidiata	Eastern short-tailed opossum	S Brazil, Uruguay, Argentina
M. domestica	Grey short-tailed opossum	E, C Brazil – N Argentina
M. emiliae		Amazon Basin; ref. 1.3
M. henseli	Hensel's short-tailed opossum	S Brazil – N Argentina
M. iheringi		S Brazil
M. kunsi		N Bolivia; lowland forest
M. maraxina		Amazon delta
M. orinoci		Venezuela, etc.; savanna
M. osgoodi (*adusta*)	Osgood's short-tailed opossum	W Bolivia, S Peru
M. scalops	Red-headed short-tailed opossum	E Brazil, N Argentina; ref. 1.4
M. sorex	Soricine short-tailed opossum	S Brazil
M. theresa (*americana*)		E Brazil
M. umbristriata		E Brazil
M. unistriata	One-striped short-tailed opossum	SE Brazil

Lestodelphys

L. halli	Patagonian opossum	S Argentina; grassland

Metachirus; (*Philander*)

M. nudicaudatus	Brown four-eyed opossum	Nicaragua – N Argentina

Didelphis; large American opossums; S, C, N America.

D. albiventris (*azarae*)		S America (mainly west)
D. marsupialis (*azarae*)		E Mexico – N Argentina, L Antilles
D. virginiana	Virginian opossum	N Costa Rica – SE Canada

Philander; (*Metachirops*)

P. mcilhennyi		E Peru; dry forest
P. opossum	Grey four-eyed opossum	E Mexico – N Argentina; forest

Virginian opossum
(Didelphis virginiana)

P. bilarni	Sandstone antechinus (Harney's marsupial mouse)	N Australia
P. macdonnellensis	Fat-tailed antechnius (Red-eared antechinus)	C Australia; rocky hills
P. ningbing		NW Australia
P. rosamondae	Little red antechinus	W Australia; grassland
P. woolleyae		W Australia; ref. 1.21

Phascogale

P. calura	Red-tailed phascogale (Red-tailed wambenger)	SW, († C, SE) Australia; dry woodland; I
P. tapoatafa	Brush-tailed phascogale (Tuan) (Common wambenger)	Australia

Dasycercus (*Dasyuroides*)

D. byrnei	Kowari	C Australia; desert, grassland
D. cristicauda	Mulgara	C Australia; desert, grassland

Dasyurus; (*Satanellus*); quolls, marsupial cats; Australia, New Guinea; forest, woodland.

D. albopunctatus	New Guinea marsupial cat	New Guinea; forest
D. geoffroii	Western quoll (Chuditch)	W († N, E) Australia, C New Guinea; woodland
D. hallucatus	Northern quoll (Satanellus)	NE, N, NW Australia
D. maculatus	Tiger quoll	E, SE Australia, Tasmania
D. spartacus	Bronze quoll	S New Guinea; ref. 1.22
D. viverrinus	Eastern Quoll	SE Australia?, Tasmania

Northern quoll
(Dasyurus hallucatus)

Sarcophilus

S. harrisii	Tasmanian devil	Tasmania; dry forest

Ningaui; Australia; desert; ref. 1.5.

N. ridei	Wongai ningaui	W, C Australia
N. timealeyi	Pilbara ningaui	NW Australia
N. yvonneae		SW, S, SE Australia

Sminthopsis; (*Antechinomys*); dunnarts; Australia, S New Guinea.

S. aitkeni (*murina*)		S Australia; ref. 1.6
S. archeri	Chestnut dunnart	New Guinea, N Queensland; ref. 1.24
S. butleri	Carpentarian dunnart	N Australia
S. crassicaudata	Fat-tailed dunnart	Australia; grassland, woodland
S. dolichura (*murina*)		SW, S Australia; ref. 1.6

S. douglasi	Julia Creek dunnart	C Queensland; I
S. fuliginosus		SW Australia
S. gilberti (*murina*)		SW Australia; ref. 1.6
S. granulipes	White-tailed dunnart	SW Australia; woodland, scrub
S. griseoventer (*murina*)		SW Australia; ref. 1.6
S. hirtipes	Hairy-footed dunnart	W, C Australia; desert
S. leucopus	White-footed dunnart	SE Australia, N Queensland, Tasmania; forest; ref. 1.23
S. longicaudata	Long-tailed dunnart	W Australia; K
S. macroura	Stripe-faced dunnart (Darling Downs dunnart)	E Australia
S. murina	Common dunnart	E Australia; forest; ref. 1.6
S. ooldea	Ooldea dunnart	S, C Australia
S. psammophila	Sandhill dunnart	C, S Australia; desert grassland; K
S. virginiae (*rufigenis*)	Red-cheeked dunnart	N Australia, S New Guinea; forest, scrub
S. youngsoni	Lesser hairy-footed dunnart	NW Australia; desert; ref. 1.7

Antechinomys; (*Sminthopsis*)

A. laniger (*spenceri*)	Kultarr (Wuhl-wuhl)	E, W, C Australia; savanna, steppe

Family Myrmecobiidae

One species.

Myrmecobius

M. fasciatus	Numbat (Banded anteater)	SW († S, SE) Australia; dry forest, woodland; E

Numbat
(Myrmerobius)

Family Thylacinidae

One species.

Thylacinus

T. cynocephalus †	Thylacine (Tasmanian wolf)	Tasmania; woodland; ? extinct

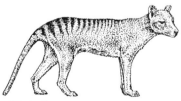

Thylacine
(Thylacinus)

Family Notoryctidae

One species.

Notoryctes

N. typhlops	Marsupial mole	S, W Australia; desert

Marsupial mole
(Notoryctes typhlops)

Family Peramelidae

Bandicoots; 18 species; Australia, New Guinea; forest, (grassland).

*Golden bandicoot
(Isoodon auratus)*

Peroryctes; New Guinea bandicoots; New Guinea.

P. broadbentii	Giant bandicoot	SE New Guinea; (in *P. raffrayanus*?)
P. longicauda	Striped bandicoot	New Guinea
P. papuensis	Papuan bandicoot	SE New Guinea
P. raffrayanus	Raffray's bandicoot	New Guinea

Microperoryctes

M. murina	Mouse-bandicoot	W New Guinea

Perameles; long-nosed bandicoots; Australia.

P. bougainville (*fasciata*)	Western Barred bandicoot (Marl)	Bernier & Dorre Is, W Australia, † mainland Australia; R
P. eremiana †	Desert bandicoot	C Australia; desert grassland; ? extinct
P. gunnii	Eastern barred bandicoot (Gunn's bandicoot)	SE Australia, Tasmania; woodland
P. nasuta	Long-nosed bandicoot	E Australia; forest, woodland

Echymipera; spiny bandicoots; New Guinea, Australia.

E. clara	White-lipped bandicoot	W New Guinea; R
E. kalubu	Spiny bandicoot	New Guinea
E. rufescens	Rufous spiny bandicoot (Rufescent bandicoot)	New Guinea, N Queensland; forest

Rhynchomeles

R. prattorum	Seram bandicoot	Seram, Indonesia

Isoodon; short-nosed bandicoots; Australia, S New Guinea; ref. 1.8.

I. arnhemensis		N Australia
I. auratus (*barrowensis*)	Golden bandicoot	W, N, († C) Australia
I. macrourus	Northern brown bandicoot (Brindled bandicoot)	NE, N Australia, S New Guinea
I. obesulus (*nauticus*) (*peninsulae*)	Southern brown bandicoot	SE, SW Australia, N Queensland, Tasmania

Chaeropus

C. ecaudatus †	Pig-footed bandicoot	SW, S Australia; woodland, scrub, grassland; ? extinct

Common rabbit bandicoot
(Macrotis lagotis)

Brushtail possum
(Trichosurus vulpecula)

Family Thylacomyidae

Rabbit-bandicoots; 2 species; Australia.

Macrotis

M. lagotis	Bilby (Common rabbit-bandicoot)	C, NW Australia; woodland, savanna; E
M. leucura †	Lesser bilby (Lesser rabbit-bandicoot)	C Australia; sandhills; ? extinct

Family Phalangeridae

Phalangers; 21 species; Australia, New Guinea, Sulawesi; forest.

Trichosurus; (*Wyulda*) Australia; forest, woodland.

T. arnhemensis	Northern brushtail possum	NW Australia
T. caninus	Mountain brushtail possum (Bobuck)	E Australia; montane forest
T. squamicaudata	Scaly-tailed possum	NW Australia; woodland; ref. 1.26
T. vulpecula	Common brushtail possum	Australia, Tasmania, [New Zealand]

Phalanger; cuscuses; N Australia; New Guinea, Sulawesi; forest, scrub.

P. carmelitae (*orientalis*)		New Guinea; montane
P. gymnotis	Aru Island ground cuscus	Aru Is (SW New Guinea)
P. interpositus (*orientalis*)	Stein's cuscus	New Guinea
P. leucippus	New Guinea ground cuscus	New Guinea
P. lullulae (*orientalis*)	Woodlark Island cuscus	Woodlark I, New Guinea; E
P. matanim		W New Guinea; montane; ref. 1.18
P. orientalis	Grey cuscus (Common phalanger)	N Queensland, New Guinea, S Moluccas, Solomon Is
P. ornatus	Moluccan cuscus	N Moluccas
P. pelengensis	Peleng Island cuscus	Peleng & Taliabu Is (E of Sulawesi)
P. permixtio		E New Guinea; ref. 1.19; R
P. rothschildi	Rothschild's cuscus	Obi I (Moluccas)
P. vestitus	Silky cuscus	New Guinea; R

Stigocuscus; (*Phalanger*).

S. celebensis	Little Celebes cuscus	Sulawesi

Potorous; potoroos; Australia; grass, scrub.

P. longipes	Long-footed potoroo	E Victoria; ref. 1.15; I
P. platyops †	Broad-faced potoroo	SW Australia; extinct
P. tridactylus (*apicalis*)	Long-nosed potoroo	E, SE, SW Australia, Tasmania

Bettongia; Australia; steppe, dry woodland

B. gaimardi (*cuniculus*)	Tasmanian bettong (Eastern bettong)	Tasmania, † E Australia
B. lesueur	Burrowing bettong (Boodie)	W Australia (small islands only), († W, C Australia); R
B. penicillata	Brush-tailed bettong (Woylie)	W, C, († S) Australia
B. tropica	Northern bettong	E Queensland; E

Aepyprymnus

A. rufescens	Rufous bettong (Rufous rat-kangaroo)	E Australia; forest, woodland

Caloprymnus

C. campestris †	Desert rat-kangaroo	C Australia; desert; ? extinct; I

Family Macropodidae

Kangaroos, wallabies; *c.* 49 species; Australia, New Guinea; steppe, savanna, forest; grazers, browsers.

Thylogale; pademelons, scrub wallabies; E Australia, New Guinea.

T. billardierii	Tasmanian pademelon (Red-bellied wallaby)	Tasmania, † SE Australia; scrub
T. brunii	Dusky wallaby	New Guinea, Bismarck Is.
T. stigmatica	Red-legged pademelon	E Australia, SE New Guinea; forest
T. thetis	Red-necked pademelon	E Australia; forest

Eastern grey kangaroo (Macropus giganteus)

Petrogale; (*Peradorcas*); rock wallabies; Australia; rocks in forest and grassland.

P. assimilis	Allied rock wallaby	N, NE Australia
P. brachyotis	Short-eared rock wallaby	N Australia
P. burbidgei	Warabi	W Australia
P. concinna	Little rock wallaby (Nabarlek)	N Australia
P. godmani	Godman's rock wallaby	Queensland
P. inornata	Unadorned rock wallaby	Queensland
P lateralis (*purpureicollis*)	Black-footed rock wallaby	Australia (except E coast)
P. penicillata	Brush-tailed rock wallaby (Western rock wallaby)	Australia, [Hawaii]
P. persephone	Prosperine rock wallaby	NE Queensland; ref. 1.10

P. rothschildi	Rothschild's rock wallaby	W Australia
P. xanthopus	Yellow-footed rock wallaby (Ring-tailed rock wallaby)	E, SE Australia

Lagorchestes; hare-wallabies; Australia; grassland.

L. conspicillatus	Spectacled hare-wallaby	N Australia
L. hirsutus	Rufous hare-wallaby (Western hare-wallaby)	W, C Australia (islands only in W); R
L. leporides †	Eastern hare-wallaby	SE Australia; extinct

Setonyx

S. brachyurus	Quokka (Short-tailed wallaby)	SW Australia; scrub

Lagostrophus

L. fasciatus	Banded hare-wallaby	Bernier & Dorre Is, W Australia, † W, S Australia; R

Macropus; (*Megaleia*, *Protemnodon*); kangaroos, wallabies; Australia, New Guinea.

M. agilis	Agile wallaby (River wallaby) (Sand wallaby)	N Australia, New Guinea; savanna
M. antelopinus	Antelope wallaroo (Antelope-kangaroo)	N Australia; savanna
M. bernardus	Black wallaroo (Bernard's wallaroo)	N Australia; rocky savanna
M. dorsalis	Black-striped wallaby	E Australia; forest
M. eugenii	Tammar wallaby (Dama wallaby)	SW, S Australia; dry forest
M. fuliginosus	Western grey kangaroo	SW, S Australia; woodland
M. giganteus	Eastern grey kangaroo	E Australia, Tasmania; woodland
M. greyi †	Toolache wallaby	S Australia, ? extinct; (in *M. irma*?)
M. irma	Western brush wallaby	SW Australia; scrub, dry forest
M. parma	Parma wallaby	SE Australia; forest, scrub
M. parryi	Whiptail wallaby (Pretty-face wallaby)	NE Australia; woodland
M. robustus	Common wallaroo (Hill kangaroo)	Australia; rocky hills in forest, grassland & desert
M. rufogriseus	Red-necked wallaby (Bennett's wallaby)	E, SE Australia, Tasmania [England, New Zealand]; scrub, woodland
M. rufus	Red kangaroo	Australia; grassland

Wallabia

W. bicolor	Swamp wallaby	E Australia; forest, thicket

Onychogalea; nail-tailed wallabies; Australia; steppe, savanna.

O. fraenata	Bridled nailtail wallaby	S Queensland, († SE Australia); E
O. lunata	Crescent nailtail wallaby	C, († SW) Australia; ? extinct
O. unguifera	Northern nailtail wallaby	N Australia

Dendrolagus; tree kangaroos; New Guinea, Queensland; forest; ref. 1.11.

D. bennettianus	Bennett's tree kangaroo	NE Queensland
D. dorianus	Unicolored tree kangaroo	New Guinea; V
D. inustus	Grizzled tree kangaroo	W, N New Guinea
D. lumholtzi	Lumholtz's tree kangaroo	NE Queensland
D. matschiei (*deltae*) (*spadix*) (*goodfellowi*)	Matschie's tree kangaroo	New Guinea; (V)
D. ursinus	Vogelkop tree kangaroo	NW New Guinea

Dorcopsis; (*Dorcopsulus*); forest wallabies; New Guinea; forest.

D. atrata	Black forest wallaby	Goodenough I, E New Guinea; R
D. hageni	Greater forest wallaby	N New Guinea
D. macleayi	Papuan forest wallaby	E New Guinea; R
D. muelleri (*veterum*)	Common forest wallaby	New Guinea
D. vanheurni	Lesser forest wallaby	New Guinea; (in *D. macleayi* ?)

Family Phascolarctidae

One species.

Phascolarctos

P. cinereus	Koala	E Australia; dry forest

*Koala
(Phascolarctos cinereus)*

Family Vombatidae

Wombats; 3 species; E, S Australia, Tasmania.

Vombatus

V. ursinus	Common wombat	E Australia, Tasmania; forest, scrub

Lasiorhinus

L. krefftii (*barnardi*) (*gillespiei*)	Northern hairy-nosed wombat	S Queensland, († SE Australia); E
L. latifrons	Southern hairy-nosed wombat	S Australia

*Southern hairy-nosed
wombat
(Lasiorhinus latifrons)*

Family Tarsipedidae

One species.

Tarsipes

T. rostratus (*spenserae*)	Honey possum	SW Australia

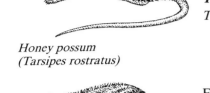

Honey possum (Tarsipes rostratus)

ORDER XENARTHRA (EDENTATA)

Edentates; *c.* 29 species; S, C, (N) America; forest – grassland; predators on invertebrates, (herbivores); ref. 2.1, 2.2.

Family Myrmecophagidae

American anteaters, vermilinguas; 4 species; S, C America; forest, savanna; terrestrial and arboreal ant-eaters.

Giant anteater (Myrmecophaga tridactyla)

Myrmecophaga

M. tridactyla	Giant anteater	N Argentina – Guatemala; forest, savanna; V

Tamandua; collared anteaters; S, C America.

T. mexicana	Northern tamandua	S Mexico – Peru, Venezuela
T. tetradactyla (*longicaudata*)	Southern tamandua	Venezuela – Uruguay

Cyclopes

C. didactylus	Pygmy anteater (Silky anteater)	S Mexico – Bolivia, N Brazil; forest

Family Bradypodidae

Three-toed sloths; 3 species; S, C America; forest; arboreal herbivores.

Bradypus; (*Scaeopus*); three-toed sloths.

Pale-throated sloth (Bradypus tridactylus)

B. torquatus	Maned sloth	SE Brazil; coastal forest; E
B. tridactylus	Pale-throated sloth	Guianas, etc.
B. variegatus (*boliviensis*) (*griseus*) (*infuscatus*)	Brown-throated sloth (Grey three-toed sloth)	Honduras – N Argentina

Family Megalonychidae

Two-toed sloths; 2 species; tropical S, C America; arboreal herbivores; includes also the extinct ground sloths.

Choloepus; two-toed sloths; formerly included in Bradypodidae or Choloepidae.

Two-toed sloth (Choloepus didactylus)

C. didactylus	Linné's two-toed sloth	Amazon Basin – Venezuela, Guianas
C. hoffmanni	Hoffmann's two-toed sloth	Nicaragua – C Brazil

Family Dasypodidae

Armadillos; *c.* 20 species; S, (C, N) America; grassland, savanna, (forest); terrestrial; mainly predators on invertebrates; ref. 2.3.

Common long-nosed armadillo (Dasypus novemcinctus)

Chaetophractus; (*Euphractus*); hairy armadillos, etc.

C. nationi	Bolivian hairy armadillo	W Bolivia, NW Argentina
C. vellerosus	Screaming armadillo (Lesser hairy armadillo)	Bolivia, Argentina
C. villosus	Larger hairy armadillo	C Argentina, Uruguay, Paraguay

Euphractus; (*Chaetophractus*)

E. sexcinctus	Yellow armadillo (Six-banded armadillo)	Uruguay – Surinam, E of Andes

Zaedyus

Z. pichiy	Pichi	S Argentina, Chile; grassland

Priodontes

P. maximus (*giganteus*)	Giant armadillo	Venezuela – N Argentina; V

Cabassous; naked-tailed armadillos; S, (C) America.

C. centralis	Northern naked-tailed armadillo	Guatemala – W Venezuela
C. chacoensis	Chacoan naked-tailed armadillo	W Paraguay, etc.
C. tatouay	Greater naked-tailed armadillo	N Argentina – S Brazil, Uruguay
C. unicinctus	Southern naked-tailed armadillo	Guyana – S Brazil

Tolypeutes; three-banded armadillos; domed armadillos.

T. matacus	Southern 3-banded armadillo	Bolivia – Argentina
T. tricinctus	Brazilian 3-banded armadillo	NE Brazil; I

Dasypus; long-nosed armadillos.

D. hybridus	Southern long-nosed armadillo	C Argentina, Uruguay – S Brazil
D. kappleri	Greater long-nosed armadillo	Peru – Surinam
D. novemcinctus	Common long-nosed armadillo (Nine-banded armadillo)	Uruguay – SE USA
D. pilosus	Hairy long-nosed armadillo	Peru; montane
D. sabanicola	Northern long-nosed armadillo	Venezuela, E Colombia

| D. septemcinctus | Brazilian long-nosed armadillo (Lesser long-nosed armadillo) | E, C Brazil – N Argentina |

Chlamyphorus; (*Calyptophractus*, *Burmeisteria*); fairy armadillos.

| C. retusus | Chacoan fairy armadillo (Greater fairy armadillo) (Burmeister's armadillo) | Bolivia – N Argentina; grassland; K |
| C. truncatus | Lesser fairy armadillo (Pink fairy armadillo) | CW Argentina; steppe; K |

ORDER INSECTIVORA

Insectivores; *c.* 365 species; Eurasia, Africa, Madagascar, N, C, (S) America; terrestrial, subterranean; predators on insects and other invertebrates.

Family Solenodontidae

Solenodons; 2 species; Cuba, Hispaniola; terrestrial omnivores.

Solenodon; (*Atopogale*).

| S. cubanus | Cuban solenodon | Cuba; E |
| S. paradoxus | Haitian solenodon | Hispaniola; E |

Haitian solenodon
(Solenodon paradoxus)

Family Tenrecidae

Tenrecs, otter-shrews; *c.* 23 species; Madagascar, W, C Africa; terrestrial, (freshwater) predators on invertebrates.

Tenrec; (*Centetes*).

| T. ecaudatus | Tail-less tenrec | Madagascar, Comoro Is; [Seychelles, Mauritius, Reunion]; dry forest |

Tail-less tenrec
(Tenrec ecaudatus)

Setifer; (*Ericulus*, *Dasogale*); ref. 3.1.

| S. setosus (*fontoynonti*) | Greater hedgehog-tenrec | Madagascar; forest, scrub |

Hemicentetes

| H. semispinosus (*nigriceps*) | Streaked tenrec | Madagascar; forest, scrub |

Echinops

| E. telfairi | Lesser hedgehog-tenrec | SW Madagascar; dry forest |

Oryzorictes; (*Nesoryctes*); mole-tenrecs (rice tenrecs); Madagascar; wet forest, fields.

O. hova		C Madagascar
O. talpoides		NW Madagascar
O. tetradactylus		C Madagascar

Microgale; (*Leptogale, Nesogale, Paramicrogale*); shrew-tenrecs; Madagascar; forest; ref. 3.2.

M. brevicaudata	Short-tailed shrew-tenrec	W, E, N Madagascar; K
M. cowani	Cowan's shrew-tenrec	E Madagascar
M. dobsoni	Dobson's shrew-tenrec	E Madagascar
M. gracilis	Long-nosed shrew-tenrec	E Madagascar; K
M. longicaudata	Lesser long-tailed shrew-tenrec	E, N Madagascar; K
M. parvula	Pygmy shrew-tenrec	N Madagascar; K
M. principula	Greater long-tailed shrew-tenrec	SE Madagascar; K
M. pulla	Dusky shrew-tenrec	NE Madagascar; ref. 3.17
M. pusilla	Lesser shrew-tenrec	S, E Madagascar
M. talazaci	Talazac's shrew-tenrec	N, E Madagascar
M. thomasi	Thomas's shrew-tenrec	E Madagascar; K

Limnogale

L. mergulus	Web-footed tenrec	Madagascar; streams; V

Geogale; (*Cryptogale*).

G. aurita		NE, SW Madagascar; K/I

Potamogale

P. velox	Giant otter-shrew	C Africa; freshwater in forest

Micropotamogale

M. lamottei	Nimba otter-shrew	Guinea, etc.; montane streams
M. ruwenzorii	Ruwenzori otter-shrew	Ruwenzori – L Kivu; montane streams

Family Chrysochloridae

Golden moles; *c.* 18 species; S Africa to Cameroun and Somalia; desert to forest; subterranean predators; taxonomy very provisional, especially at generic level.

Chrysochloris

C. asiatica	Cape golden mole	W Cape Province
C. stuhlmanni (*fosteri*)	Stuhlmann's golden mole	E Africa
C. visagiei	Visagie's golden mole	W Cape Province; K

Giant golden mole
(*Chrysospalax trevelyani*)

Eremitalpa

E. granti	Grant's golden mole	SW Africa; coastal desert

Calcochloris; (*Amblysomus*).

C. obtusirostris	Yellow golden mole	S Africa

Cryptochloris

C. wintoni	De Winton's golden mole	W Cape Province; desert; K
C. zyli	Van Zyl's golden mole	SW Cape Province; K

Amblysomus

A. gunningi	Gunning's golden mole	Transvaal; montane forest; K
A. hottentotus	Hottentot golden mole	S Africa
A. iris	Zulu golden mole	SE Africa
A. julianae	Juliana's golden mole	Transvaal; R

Chlorotalpa

C. arendsi	Arends's golden mole	E Zimbabwe; montane
C. duthiae	Duthie's golden mole	S Cape Province; R
C. leucorhina	Congo golden mole	WC Africa
C. sclateri	Sclater's golden mole	S Africa
C. tytonis	Somali golden mole	Somalia

Chrysospalax

C. trevelyani	Giant golden mole	SE Cape Province; forest; E
C. villosus	Rough-haired golden mole	SE Africa; grassland; K

Family Erinaceidae

Hedgehogs, moonrats; *c.* 19 species; Eurasia (except boreal zones), Africa; forest to desert; terrestrial predators on invertebrates; ref. 3.18.

Subfamily Galericinae (Echinosoricinae)

Moonrats, gymnures; 6 species; SE Asia; forest.

Echinosorex; (*Gymnurus*).

E. gymnurus	Moonrat	Malaya, Sumatra, Borneo; forest

Hylomys

Lesser moonrat
(Hylomys suillus)

H. suillus	Lesser moonrat	Yunnan – Java, Borneo

Neotetracus; (*Hylomys*)

N. sinensis	Shrew-hedgehog	S China, Indochina

Neohylomys; (*Hylomys*)

N. hainanensis	Hainan moonrat	Hainan I; China

Podogymnura

P. aureospinula	Spiny moonrat	Dinagat I, Philippines; ref. 3.3
P. truei	Mindanao moonrat	Mindanao, Philippines; forest; V

Subfamily Erinaceinae

Spiny hedgehogs; 13 species; Europe, Asia (temperate and SW), Africa.

Erinaceus; woodland hedgehogs; Europe, N Asia.

European hedgehog
(Erinaceus europaeus)

E. amurensis	Amur hedgehog	Amur – Korea, NE, E China

| *E. concolor* | E European hedgehog | E Europe – Syria, W Siberia |
| *E. europaeus* | W European hedgehog | W, N Europe, [New Zealand] |

Atelerix; (*Erinaceus, Aethechinus*); African hedgehogs; Africa; mainly savanna.

A. albiventris	Four-toed hedgehog	Senegal – Kenya – Zambezi; savanna
A. algirus	Algerian hedgehog	NW Africa, Canary Is, [SW Europe]; scrub, steppe
A. frontalis	S African hedgehog	S Africa – Zambezi
A. sclateri	Somali hedgehog	N Somalia

Hemiechinus; steppe hedgehogs; W, C Asia.

H. auritus	Long-eared hedgehog	Libya – Mongolia & Pakistan
H. collaris	Indian long-eared hedgehog	NW India, Pakistan
H. dauuricus (*hughi*) (*sylvaticus*)	Daurian hedgehog	E of Gobi Desert – Shanxi, China; dry steppe

Paraechinus; desert hedgehogs; Sahara to India.

P. aethiopicus (*deserti*) (*dorsalis*)	Desert hedgehog	Sahara, Arabia; desert
P. hypomelas	Brandt's hedgehog	Iran, Turkestan – Pakistan; desert, dry steppe
P. micropus (*nudiventris*)	Indian hedgehog	NW, SW India

Family Soricidae

Shrews; *c.* 272 species; Eurasia, Africa, N America to northern S America; forest to desert; terrestrial (or partially aquatic) insectivores.

Eurasian common shrew (Sorex araneus)

Sorex; (*Microsorex*); red-toothed shrews; N America, N Eurasia; tundra, forest; ref. 3.4 (N America); 3.19 (China).

S. alaskanus	Glacier Bay water shrew	S Alaska; (in *S. palustris*?)
S. alpinus	Alpine shrew	Europe; montane forest
S. araneus	Eurasian common shrew	Europe – R Yenesei; forest, tundra
S. arcticus	Arctic shrew (Black-backed shrew)	Alaska, Canada, N USA; wet forest
S. arizonae	Arizona shrew	SE Arizona, etc.; montane; close to *S. ventralis*
S. asper	Tien Shan shrew	C Asia; montane
S. bedfordiae	Lesser striped shrew	Gansu – Nepal; montane forest

S. bendirii	Pacific water shrew	W coast USA; wet forest
S. caecutiens (*annexus*)	Laxmann's shrew	E Europe – Japan; con. forest, tundra
S. cansulus		Gansu, China
S. cinereus	Masked shrew (American common shrew)	N USA, Canada, NE Siberia; forest, tundra
S. coronatus		France, etc.; close to *S. araneus*
S. cylindricauda	Greater striped shrew	Sichuan, China; montane forest
S. daphaenodon	Large-toothed Siberian shrew	Siberia, NE China; con. forest, tundra
S. dispar	Long-tailed shrew	NE USA
S. emarginatus		Mexico; close to *S. ventralis*
S. excelsus		N Yunnan – Sichuan
S. fontinalis	Maryland shrew	E USA
S. fumeus	Smoky shrew	SE Canada, NE USA
S. gaspensis	Gaspé shrew	Gaspé Peninsula, Quebec; close to *S. dispar*
S. gracillimus	Slender shrew	NE Asia, N Japan
S. granarius		Spain; close to *S. araneus*
S. haydeni	Prairie shrew	N USA, S Canada; close to *S. cinereus*
S. hosonoi	Azumi shrew	C Honshu, Japan; montane
S. hoyi (*thompsoni*)	American pygmy shrew	Canada, NE USA; con. forest; (formerly in *Microsorex*)
S. hydrodromus	Pribilof shrew	St Paul I, Alaska
S. isodon (*sinalis*)		NE Europe – SE Siberia; con. forest
S. jacksoni (? *pribilofensis*) (*ugyunak*)	Barrenground shrew (St Lawrence Island shrew)	Alaska; ? N Canada, NE Siberia
S. longirostris	Southeastern shrew	SE USA
S. lyelli	Mount Lyell shrew	California; montane
S. macrodon	Large-toothed shrew	Veracruz, Mexico; forest
S. merriami	Merriam's shrew	W USA; high steppe
S. milleri	Carmen Mountain shrew	NE Mexico; montane
S. minutissimus	Least shrew	Scandinavia, Siberia, Korea, S China, Japan; con. forest
S. minutus	Eurasian pygmy shrew	Europe – C Siberia, ? Tibet, C China; forest, tundra
S. mirabilis	Giant shrew	E Siberia, N Korea

S. monticolus (obscurus)	Dusky shrew	Alaska, W Canada, W USA, NW Mexico; tundra, forest
S. nanus	Dwarf shrew	Wyoming – New Mexico; montane
S. oreopolis	Mexican long-tailed shrew	Mexico; montane
S. ornatus (juncensis) (sinuosus)	Ornate shrew	California, NW Mexico
S. pacificus	Pacific shrew	W USA; coastal forest
S. palustris	American water shrew	Canada, USA; wet forest
S. preblei	Preble's shrew	E Oregon – Montana, USA
S. raddei	Radde's shrew	Caucasus, etc.
S. roboratus (vir)	Flat-skulled shrew	E Siberia – Altai
S. sadonis		Sado Is, Japan; ref. 3.21
S. samniticus (araneus)	Apennine shrew	S, C Italy
S. satunini	Caucasian shrew	Caucasus etc., ref. 3.20
S. saussurei	Saussure's shrew	C Mexico – Guatemala; montane
S. sclateri	Sclater's shrew	Chiapas, SE Mexico
S. sinalis (isodon)		N Europe – N China; con. forest
S. stizodon	San Cristobal shrew	Chiapas, SE Mexico
S. tenellus	Inyo shrew	California, Nevada
S. thibetanus (buchariensis) (kozlovi) (planiceps)		Pamirs – W China; montane
S. trowbridgii	Trowbridge shrew	W coast USA; forest
S. tundrensis (arcticus) (centralis)	Tundra shrew	Alaska, NE Siberia
S. unguiculatus	Long-clawed shrew	E Siberia, N Japan
S. vagrans (trigonirostris)	Vagrant shrew (Wandering shrew)	W USA, S Mexico; marsh, wet forest
S. veraepacis	Verapaz shrew	SE Mexico; montane
S. ventralis		Mexico
S. volnuchini (minutus)	Caucasian pygmy shrew	Caucasus

Soriculus; (*Chodsigoa*, *Episoriculus*); mountain shrews; Himalayas, China; montane forest; ref. 3.5.

S. caudatus	Hodgson's brown-toothed shrew	Himalayas – S China, Burma
S. fumidus (caudatus)	Taiwan brown-toothed shrew	Taiwan; close to *S. caudatus*
S. hypsibius	De Winton's shrew	SW, C China

S. lamula (*hypsibius*)		Yunnan – Gansu

S. leucops
  (*baileyi*)
  (*gruberi*)
	Indian long-tailed shrew	C Nepal – S China –
		  N Vietnam

S. macrurus (*leucops*)		C Nepal – Vietnam
S. nigrescens	Himalayan shrew	Himalayas – Yunnan

S. parca
  (*smithii*)
  (*lowei*)
		Thailand – Sichuan

S. salenskii	Salenski's shrew	N Sichuan; in *S. smithii?*
S. smithii	Smith's shrew	Sichuan – Shaanxi

Neomys; Eurasian water shrews; N Eurasia; wet forest, streams, marshes.

N. anomalus	Southern water shrew	S, E Europe
N. fodiens	Eurasian water shrew	W Europe – E Siberia, Korea
N. schelkovnikovi	Transcaucasian water shrew	Armenia, Georgia

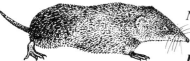

Northern short-tailed shrew
(Blarina brevicanda)

Blarina; American short-tailed shrews; E N America; forest, grassland.

B. brevicauda	Northern short-tailed shrew	E USA, SE Canada
B. carolinensis (*telmalestes*)	Southern short-tailed shrew	SE USA
B. hylophaga (*carolinensis*)	Elliot's short-tailed shrew	S, SE USA

Blarinella

B. quadraticauda	Chinese short-tailed shrew	C, S China; montane forest

Cryptotis; small-eared shrews; E USA – Ecuador, Surinam; forest, (grassland).

C. avia		Colombia
C. endersi	Ender's small-eared shrew	W Panama

C. goldmani	Goldman's small-eared
	  shrew
		S Mexico; wet montane
		  forest

C. goodwini	Goodwin's small-eared
	  shrew
		S Guatemala, S Mexico;
		  montane forest

C. gracilis	Talamancan small-eared shrew	Honduras – Panama; forest
C. magna	Big small-eared shrew	Oaxaca, S Mexico; montane forest

C. mexicana
  (*phillipsi*)
	Mexican small-eared shrew	S, E Mexico; wet montane
		  forest

C. montivaga		S Ecuador
C. nigrescens	Blackish small-eared shrew	S Mexico – Panama; scrub, forest
C. parva	American least shrew	C, E USA – Panama; grassland, scrub, forest

C. squamipes		W Colombia
C. thomasi		Ecuador – Venezuela

Notiosorex

N. crawfordi	Desert shrew (Grey shrew)	SW USA, Mexico; semidesert scrub

Megasorex; (*Notiosorex*).

M. gigas	Giant Mexican shrew (Merriam's desert shrew)	SW Mexico

Crocidura; (*Praesorex*); white-toothed shrews; Eurasia, Africa; forest to semi-desert; taxonomy very provisional in some areas; ref. 3.6 (Oriental Region).

(*Crocidura suaveoleus*)

C. allex		Kenya, N Tanzania
C. andamanensis		S Andaman I, Indian Ocean
C. ansellorum		NW Zambia; ref. 3.22, 23
C. arabica		S Arabia; ref. 3.24
C. attenuata (*aequicauda*) (*trichura*)		Himalayas – S China – Java, Taiwan, Baton Is (N Philippines)
C. baileyi		Ethiopia; montane
C. baluensis		Borneo, ? Sumatra; montane
C. beatus		Mindanao, Philippines
C. bloyeti		C Tanzania
C. bottegi		W Africa, Ethiopia, N Kenya
C. bovei		Zaire
C. butleri		Sudan, Uganda, Kenya, S Somalia
C. buettikoferi		W Africa; forest
C. caliginea		NE Zaire; forest
C. canariensis	Canary shrew	Canary Is; ref. 3.25
C. cinderella		Gambia, ? Mali
C. congobelgica		NE Zaire
C. crenata		Gabon
C. crossei		Nigeria – Ivory Coast; forest
C. cyanea	Reddish-grey musk shrew	Ethiopia – S Africa
C. denti		C Africa
C. dolichura		Guinea – Uganda
C. douceti		Ivory Coast, Guinea
C. dsinezumi		Japan except Hokkaido, ? Taiwan
C. eisentrauti		Mt Cameroun
C. elgonius		W Kenya, N Tanzania
C. elongata		Sulawesi
C. erica		W Angola
C. fischeri		S Ethiopia – Zaire

C. flavescens	Greater musk shrew	S Africa
C. floweri	Flower's shrew	Egypt
C. foxi		Nigeria – Ghana; Guinea savanna
C. fuliginosa (malayana) (palawanensis)		S China – Malaya, Sumatra, Java, Borneo, Palawan, ? Timor
C. fulvastra (arethusa) (sericea)		Mali – Ethiopia, Kenya; ref. 3.8
C. fumosa		E Africa; montane forest
C. fuscomurina (bicolor)	Tiny musk shrew	Uganda – N Cape Prov., Namibia; savanna; ref. 3.9
C. glassi		Harar, Ethiopia; montane
C. gracilipes (hildegardeae)	Peters' musk shrew	W, E S Africa
C. grandiceps		Ivory Coast – S Nigeria; ref. 3.11
C. grandis		Mindanao, Philippines
C. grassei		C Africa
C. grayi (halconis)		Luzon, Mindoro, Philippines
C. greenwoodi		S Somalia
C. gueldenstaedtii (? cypria) (? vorax)		Caucasus – Israel, Crete, ? Cyprus; ? – SW China
C. hirta	Lesser red musk shrew	S Africa – S Somalia
C. hispida	Andaman spiny shrew	Middle Andaman I, Indian Ocean
C. horsfieldii		Sri Lanka, Indochina, Hainan, Taiwan, Ryukyu Is
C. jacksoni		E Africa
C. kivuana		Kivu, E Zaire
C. lamottei		Ivory Coast, Togo
C. lanosa		E Zaire, Rwanda
C. lasia		Asia Minor, etc.
C. lasiura		Korea – Ussuri
C. latona		E Zaire
C. lea		Sulawesi
C. leucodon		C, S Europe – Israel
C. levicula		Sulawesi
C. littoralis		Uganda – Cameroun
C. longipes		W Nigeria; swamps; ref. 3.12
C. lucina		Ethiopia; montane; ref. 3.16
C. luluae		S Zaire

C. luna	Greater grey-brown musk shrew	E Zimbabwe – Kenya; savanna
C. lusitania		W Sahara
C. macarthuri		Kenya
C. macowi		Mt Nyiro, N Kenya
C. manengubae		Cameroun; ref. 3.7
C. maquassiensis	Makwassie musk shrew	Transvaal, Zimbabwe; R
C. mariquensis	Swamp musk shrew	S Africa – Zambia
C. maurisca		Uganda, Kenya, Tanzania
C. mindorus		Mindoro, Philippines
C. miya	Ceylon long-tailed shrew	Sri Lanka
C. monax		E Africa; montane
C. monticola (*bartelsii*)		Malaya, Borneo, Java – Flores
C. montis		E Africa; montane; ref. 3.27
C. mutesae		Uganda
C. nana		Somalia, Ethiopia; (in *C. religiosa*?)
C. nanilla		W, C, E Africa
C. neglecta		Sumatra, Java etc.
C. negrina		Negros I, Philippines
C. nicobarica (*jenkinsi*)		Nicobar & Andaman Is, Indian Ocean
C. nigricans		Angola
C. nigripes		Sulawesi
C. nigrofusca		Zaire
C. nimbae		Ivory Coast – Guinea
C. niobe		Ruwenzori, Uganda, SW Ethiopia
C. occidentalis (*olivieri*) (? *bolivari*)		Senegal – Egypt – Zimbabwe
C. odorata (*goliath*)		Guinea – Gabon
C. orientalis		Java; montane
C. orii		Ryukyu Is.
C. osorio		Gran Canaria, Canary Is; ref. 3.35
C. paradoxura		Sumatra, ? Java
C. pasha		Sudan – Mali; savanna
C. pergrisea (? *armenica*)		Asia Minor – Kashmir; montane
C. phaeura		S Ethiopia
C. picea		Cameroun
C. pitmani		Zambia
C. planiceps		N Uganda, S Sudan
C. poensis		W, C Africa

C. raineyi		Kenya; ref. 3.27
C. religiosa	Egyptian pygmy shrew	Egypt
C. rhoditis		Sulawesi
C. roosevelti		Uganda, Tanzania – Angola
C. russula (*foucauldi*)		C, S Europe, N Africa; ref. 3.28
C. sansibarica		Zanzibar, Pemba I
C. selina		Uganda; ref. 3.27
C. serezkyensis		Pamir Mts, etc.; ref. 3.15
C. sibirica		C Asia
C. sicula		Sicily
C. silacea	Lesser grey-brown musk shrew	Transvaal – Zimbabwe, ? – Kenya
C. smithi		Ethiopia, Somalia, Senegal
C. somalica		Sudan, Somalia, Ethiopia, Oman
C. suaveolens		SW Europe, Arabia – Korea, China, Taiwan
C. susiana		SW Iran; steppe
C. tansaniana		Usambara Mts; Tanzania; ref. 3.31
C. tarfayensis		Morocco; ref. 3.28
C. telfordi		Uluguru Mts, Tanzania; ref. 3.31
C. tenuis		Timor
C. thalia		Ethiopia; montane; 3.16
C. theresae		Guinea – Ghana
C. thomensis		Sao Tome I, W Africa; ref. 3.32
C. turba		Zambia – N Angola
C. usambarae		Usambara Mts, Tanzania; ref. 3.16
C. viaria (*suahelae*) (*bolivari*) (*tephra*)		Morocco – Senegal – Kenya; ref. 3.8
C. vulcani		Mt Cameroun
C. whitakeri		NW Africa
C. wimmeri		Liberia – Gabon
C. xantippe		Kenya, Tanzania
C. yankariensis		N Nigeria, Sudan; ref. 3.10
C. zaodon		Zambia – S Sudan, ? W Africa
C. zaphiri		Kenya, S Ethiopia
C. zarudnyi		Afghanistan, Baluchistan
C. zimmeri		S Zaire
C. zimmermanni		Crete; ref. 3.33

Suncus; (*Pachyura*, *Podihik*); Africa, S Eurasia; forest, scrub, savanna.

S. ater	Black shrew	Mt Kinabalu, Borneo
S. dayi		S India
S. etruscus	Pygmy white-toothed shrew	Mediterranean – India, Sri Lanka; W Africa; scrub, etc.
S. hosei		Sarawak, Borneo
S. infinitesimus	Least dwarf shrew	S Africa – Kenya, ? W Africa
S. lixus	Greater dwarf shrew	Kenya – Angola, Transvaal
S. malayanus		Malaya, Borneo; (in *S. etruscus*?)
S. mertensi		Flores
S. montanus (*murinus*)		S India, Sri Lanka
S. murinus (*luzoniensis*) (*occultidens*) (*palawanensis*)	House shrew	S Asia, (E Africa); commensal
S. remyi		Gabon
S. stoliczkanus		India, etc.
S. varilla	Lesser dwarf shrew	S Africa – Tanzania, Zaire
S. zeylanicus (*murinus*)		Sri Lanka

Feroculus

F. feroculus	Kelaart's long-clawed shrew	Sri Lanka; montane forest

Solisorex

S. pearsoni	Pearson's long-clawed shrew	Sri Lanka; montane

Paracrocidura; ref. 3.30

P. graueri		E Zaire
P. maxima		Ruanda – Ruwenzori; montane
P. schoutedeni		Cameroun – Zaire

Sylvisorex; (*Suncus*); Africa, forest, grassland.

S. granti		Cameroun – E Africa; montane
S. howelli		Uluguru Mts, Tanzania; ref. 3.13
S. johnstoni		Cameroun, Gabon, Zaire, Bioco
S. lunaris		C Africa; montane
S. megalura	Climbing shrew	Guinea – Ethiopia – Zimbabwe
S. morio		Mt Cameroun, Bioco

House shrew
(Suncus murinus)

S. ollula		Cameroun – Zaire
S. vulcanorum		Ruanda etc.; montane forest; ref. 3.34

Ruwenzorisorex; (*Sylvisorex*)

R. suncoides		C Africa; montane; ref. 3.29

Myosorex; (*Surdisorex*); mouse-shrews; C, S Africa; forest.

M. baboulti		E Zaire
M. blarina		C Africa; montane
M. cafer	Dark-footed forest shrew	S Africa, Zimbabwe
M. eisentrauti		Cameroun, Bioco
M. geata		Tanzania
M. longicaudatus	Long-tailed forest shrew	Knysna, S Africa
M. norae		Aberdare Mts, Kenya
M. polli		Kasai, Zaire
M. polulus		Mt Kenya; montane scrub
M. preussi		Mt Cameroun
M. varius	Forest shrew	S Africa

Diplomesodon

D. pulchellum	Piebald shrew	Russian Turkestan; desert

Anourosorex

A. squamipes	Mole-shrew	C China – N Thailand, Taiwan; montane forest

Chimarrogale; (*Crossogale*); oriental water shrews; E Asia; montane streams; ref. 3.19.

C. himalayica		Himalayas, China, Taiwan
C. phaeura		Malaya, Sumatra, Borneo
C. platycephala		Japan
C. styani		Sichuan, N Burma

Nectogale

N. elegans	Elegant water shrew	Sikkim – Shaanxi; montane streams

Scutisorex

S. somereni (*congicus*)	Armoured shrew	C Africa; forest

Family Talpidae

Moles, shrew-moles, desmans; *c.* 31 species; Eurasia, N America; mainly subterranean in forest and wet grassland, some aquatic; predators on invertebrates.

Subfamily Uropsilinae

Chinese shrew-moles; 3 species; S China, etc.; montane; ref. 3.14.

Uropsilus; (*Nasillus, Rhynchonax*).

U. andersoni		C Sichuan
U. gracilis		Sichuan, Yunnan, N Burma
U. soricipes		C Sichuan

Subfamily Desmaninae

Desmans; 2 species; Europe; aquatic.

Desmana

D. moschata	Russian desman	European Russia, [C Siberia]; rivers; V

Pyrenean desman
(Galemys pyrenaicus)

Galemys

G. pyrenaicus	Pyrenean desman	Pyrenees, Iberia; rivers; V

Subfamily Talpinae

Moles; *c.* 26 species; Eurasia, N America; subterranean; generic and specific classification of Eurasian forms very provisional.

Talpa; (*Mogera, Parascaptor, Scaptochirus, Euroscaptor*); Eurasian moles; Eurasia; subterranean in forest and grassland.

T. altaica	Siberian mole	W, C Siberia
T. caeca	Mediterranean mole	S Europe – Caucasus
T. caucasica	Caucasian mole	Caucasus
T. europaea	European mole	Europe, W Siberia
T. grandis (*micrura*)		Sichuan, Yunnan
T. insularis (*latouchei*)		SE China, Hainan, Taiwan
T. leucura		Yunnan, Burma, Assam
T. longirostris (*micrura*)		S China
T. micrura (*klossi*)	Himalayan mole	E Himalayas – Malaya
T. mizura	Japanese mountain mole	Honshu, Japan
T. moschata	Short-faced mole	E China
T. robusta (*kobeae*)	Large Japanese mole	Korea, etc. – Japan
T. romana	Roman mole	Italy, Balkans
T. streeti	Persian mole	NW Iran
T. wogura (*coreana*)	Japanese mole	Japan, ? Korea, etc.

European mole
(Talpa europaea)

Scaptonyx

S. fusicaudus	Long-tailed mole	S China, N Burma; montane

Neurotrichus

N. gibbsii	American shrew-mole	W coast USA; wet forest

Urotrichus; (*Dymecodon*).

U. pilirostris	Lesser Japanese shrew-mole	Japan; montane forest
U. talpoides	Greater Japanese shrew-mole	Japan

Scapanulus

S. oweni	Kansu mole	C China; montane forest

Parascalops

P. breweri	Hairy-tailed mole	NE USA, SE Canada

Scapanus; western moles; W USA; wet forest, grassland.

S. latimanus	Broad-footed mole	California, etc.
S. orarius	Coast mole	W USA
S. townsendii	Townsend's mole	W coast USA

Scalopus

S. aquaticus (*inflatus*) (*montanus*)	Eastern American mole	C, E USA, NE Mexico

Condylura

C. cristata	Star-nosed mole	NE USA, SE Canada

ORDER SCANDENTIA

One family, variously included in the Insectivora or Primates, or associated with the Macroscelidea, but probably better treated as a separate order.

Family Tupaiidae

Tree shrews; *c.* 16 species; SE Asia; forest; arboreal insectivores.

Common tree shrew (Tupaia glis)

Tupaia

T. belangeri (*glis*)	Northern tree shrew	Nepal – S China – Thailand
T. dorsalis	Striped tree shrew	Borneo
T. glis (*palawanensis*)	Common tree shrew	Malaya, Sumatra, Java, Borneo, Palawan
T. gracilis	Slender tree shrew	Borneo
T. javanica	Javan tree shrew	Java, Sumatra
T. minor	Pygmy tree shrew	Malaya, Sumatra, Borneo
T. montana	Mountain tree shrew	Borneo; montane
T. nicobarica	Nicobar tree shrew	Nicobar Is, Indian Ocean
T. picta	Painted tree shrew	Borneo
T. splendidula	Ruddy tree shrew (Red-tailed tree shrew)	S Borneo; lowland
T. tana	Large tree shrew	Borneo, Sumatra

Anathana

A. ellioti	Madras tree shrew	S, C India

Dendrogale
D. melanura	Bornean smooth-tailed tree shrew	Borneo; montane
D. murina	Northern smooth-tailed tree shrew	S Indochina

Urogale
U. everetti	Philippine tree shrew	Mindanao, etc., Philippines

Ptilocercus
P. lowii	Pen-tailed tree shrew	Malaya, Sumatra, W, N Borneo

ORDER DERMOPTERA

One family.

Family Cynocephalidae

Flying lemurs (colugos); 2 species; SE Asia; forest; arboreal, gliding, herbivores.

Cynocephalus; (*Galeopithecus, Galeopterus*).
C. variegatus	Malayan flying lemur	Indochina – Java, Borneo
C. volans	Philippine flying lemur	S Philippines

Malayan flying lemur
(Cynocephalus variegatus)

ORDER CHIROPTERA

Bats; *c.* 977 species; worldwide.

SUBORDER MEGACHIROPTERA

Family Pteropodidae

Fruit bats, flying foxes; *c.* 162 species; Old-world tropics and subtropics: Africa, S Asia to Australia, W Pacific Islands; mainly forest, rarely in caves; frugivorous, nectarivorous; ref. 4.9.

Subfamily Pteropodinae

Eidolon
E. helvum	Straw-coloured fruit bat	Africa S of Sahara, Ethiopia, SW Arabia, Madagascar

Egyptian rousette
(Rousettus aegyptiacus)

Rousettus; rousettes; Africa, S Asia – Solomon Is.
Subgenus *Rousettus*
R. aegyptiacus	Egyptian rousette	S Africa – Senegal, Ethiopia, Egypt – Lebanon – Pakistan, Cyprus

R. amplexicaudatus (*stresemanni*)	Geoffroy's rousette	S Burma – Solomon Is, Philippines; refs. 4.10, 11, 12
R. celebensis	Sulawesi rousette	Sulawesi, Sanghir Is, (?) Talaud Is.
R. lanosus	Ruwenzori long-haired rousette	S Ethiopia – Tanzania, E Zaire
R. leschenaulti	Leschenault's rousette	Pakistan – Vietnam – S China; Sri Lanka, Java, Bali
R. madagascariensis	Madagascar rousette	Madagascar
R. obliviosus		Comoro Is
R. spinalatus		N Sumatra, Borneo; refs. 4.11, 13, 14

Subgenus *Boneia*

R. bidens		N Sulawesi; ref. 4.110

Subgenus *Lissonycteris*

R. angolensis	Bocage's fruit bat	Guinea – Ethiopia, Angola, Zimbabwe, Zambia, Mozambique

Myonycteris

M. brachycephala	Sào Tomé collared fruit bat	Saò Tomé I, Gulf of Guinea
M. relicta		SE Kenya, NE Tanzania; ref. 4.15
M. torquata	Little collared fruit bat	Sierra Leone – Angola, Zambia

Pteropus; flying foxes; Madagascar, SE Asia, N Australia, islands of Indian Ocean and W Pacific.

P. admiralitatum	Admiralty flying fox	Admiralty Is – Solomons
P. alecto	Central flying fox (Black flying fox)	Sulawesi – S New Guinea, NW, N, NE Australia
P. anetianus		Vanuatu, Banks Is
P. argentatus	Silvery flying fox	? Amboina I; ref. 4.16
P. brunneus		Percy I, off E Queensland; perhaps applied to a vagrant *P. hypomelanus*
P. caniceps	Ashy-headed flying fox	Sulawesi, N Moluccas, etc.
P. chrysoproctus	Amboina flying fox	S Moluccas; ref. 4.110
P. conspicillatus	Spectacled flying fox	N Moluccas, New Guinea – NE Queensland
P. dasymallus	Ryukyu flying fox	S Japan, Ryukyu Is, Taiwan
P. faunulus		Nicobar Is
P. fundatus		Banks Is, Vanuatu
P. giganteus	Indian flying fox	India – Burma, SW China, Sri Lanka, Maldive Is, Andaman Is

Seychelles flying fox
(*Pteropus seychellensis*)

P. gilliardi	Gilliard's flying fox	New Britain, Bismarck Arch.
P. griseus	Grey flying fox	Timor – Sulawesi, ? Luzon
P. howensis		Ontong Java Atoll, Solomons
P. hypomelanus (*? santacrucis*) (*? mearnsi*)	Small flying fox	S Burma – Solomons, Philippines, C Maldive Is
P. insularis		Caroline Is; I
P. intermedius		S Burma
P. leucopterus		Philippines
P. livingstonii	Comoro black flying fox	Anjouan I, Moheli I, Comoro Is; ref. 4.17; E
P. lombocensis	Lombok flying fox	Lesser Sunda Is
P. lylei	Lyle's flying fox	Thailand, Indochina
P. macrotis (*pohlei*)	Big-eared flying fox	S New Guinea, Aru Is; I
P. mahaganus	Lesser flying fox	Ysabel I, Bougainville I, Solomons
P. mariannus	Marianas flying fox	Mariana Is, Palau Is, Caroline Is, Ryukyu Is; V
P. melanopogon	Black-bearded flying fox	S Moluccas; ref. 4.110
P. melanotus (*satyrus*)		Andaman Is, Nicobar Is – Christmas I, Indian Ocean
P. molossinus		E Caroline Is; I
P. neohibernicus (*sepikensis*)	Bismarck flying fox	New Guinea, Bismarck Arch.
P. niger	Greater Mascarene flying fox	† Reunion, Mauritius; ref. 4.17; V
P. nitendiensis		Santa Cruz Is, SW Pacific
P. ocularis	Ceram flying fox	Buru, Ceram Is, S Moluccas
P. ornatus		New Caledonia; Loyalty Is
P. personatus	Masked flying fox	N Moluccas, (?) Sulawesi; ref. 4.110
P. phaeocephalus		Mortlock I (= Tauu I), C Caroline Is; I
P. pilosus†		Palau Is; extinct
P. poliocephalus	Grey-headed flying fox	E Australia, vagrant Tasmania
P. pselaphon		Bonin Is; Volcano Is, (S of Japan)
P. pumilus (*balutus*) (*tablasi*)		Philippines; ref. 4.18
P. rayneri (*cognatus*)	Solomon flying fox	Solomon Is

P. rodricensis	Rodriguez flying fox	Rodriguez I; † Round I; ref. 4.17; E
P. rufus	Madagascar flying fox	Madagascar
P. samoensis	Samoa flying fox	Fiji Is, Samoa Is; E
P. scapulatus	Little red flying fox	W, N, E Australia; S New Guinea; vagrant New Zealand
P. seychellensis	Seychelles flying fox	Comoro Is, Aldabra I, Seychelles; Mafia I; (V)
P. speciosus		Philippines, etc. (in *P. griseus*?)
P. subniger†	Lesser Mascarene flying fox	Reunion, Mauritius; extinct; ref. 4.17
P. temmincki	Temminck's flying fox	S Moluccas, Bismarck Arch.
P. tokudae†	Guam flying fox	Guam I; extinct
P. tonganus	Insular flying fox	Karkar I (NE New Guinea) – Samoa, Cook Is; I
P. tuberculatus		Vanikoro I, Santa Cruz Is
P. vampyrus	Large flying fox	S Burma – Java, Philippines, Borneo, Timor
? *P. vanikorensis*		Vanikoro Is, Santa Cruz Is
P. vetulus		New Caledonia
P. voeltzkowi	Pemba flying fox	Pemba I, off Tanzania; V
P. woodfordi	Least flying fox	Solomons

Pteralopex

P. acrodonta		Taveuni I, Fiji Is
P. anceps		Bougainville, Choiseul Is, Solomons
P. atrata	Cusp-toothed flying fox	Ysabel, Guadalcanal Is, Solomons

Acerodon; Philippines – Lesser Sundas.

A. celebensis (*arquatus*)	Sulawesi flying fox	Sulawesi; ref. 4.16
A. humilis	Talaud flying fox	Talaud Is, N Moluccas
A. jubatus		Philippines
A. leucotis		Busuanga I, Palawan, Balabac, Philippines; ref. 4.16
A. lucifer		Panay I, C Philippines; extinct ?
A. macklotii	Sunda flying fox	Lesser Sunda Is

Neopteryx

N. frosti	Small-toothed fruit bat	N, W Sulawesi

Styloctenium
S. wallacei   Striped-faced fruit bat   Sulawesi

Dobsonia; naked-backed fruit bats; Philippines – Australia.
D. beauforti   Biak I, Owii I, Waigeo I,
                                                      off NW New Guinea
D. chapmani   Cebu I, Negros I,
                                                      Philippines
D. emersa   Biak I, Owii I, off NW
                                                      New Guinea; ref. 4.19
D. exoleta   Sulawesi naked-backed bat   Sulawesi
D. inermis   Solomons naked-backed   Solomons
                       bat
D. minor   Lesser naked-backed bat   C Sulawesi, C, W New
                                                      Guinea; ref. 4.110, 113;
                                                      K
D. moluccense   Greater naked-backed bat   Moluccas, New Guinea
  (? anderseni)
D. pannietensis   Bismarck Arch.,
  (anderseni)   Trobriand,
  (? remota)   D'Entrecasteaux,
                                                      Louisiade Is, N
                                                      Queensland; ref. 4.19
D. peronii   Western naked-backed bat   Lesser Sunda Is, Timor
D. praedatrix   New Britain naked-backed   Bismarck Archipelago
                       bat
D. viridis   Greenish naked-backed bat   Sulawesi; Moluccas; ref.
  (? crenulata)   4.12

Plerotes; ref. 4.111
P. anchietae   Anchieta's fruit bat   Angola, S Zaire, Zambia

Hypsignathus
H. monstrosus   Hammer-headed fruit bat   Gambia – Ethiopia –
                                                      Zaire, NE Angola

Hammer-headed fruit bat
(Hypsignathus monstrosus)

Epomops; epauletted fruit bat; ref. 4.111.
E. buettikoferi   Büttikofer's fruit bat   Guinea – Ghana;
                                                      ? Nigeria
E. dobsonii   Dobson's fruit bat   W, C Angola – S Zaire,
                                                      Rwanda, NE Botswana,
                                                      Zambia, Tanzania
E. franqueti   Franquet's fruit bat   Ivory Coast – S Sudan –
                       (Singing fruit bat)   NW Tanzania, Angola

Epomophorus; epauletted fruit bats; Africa S of Sahara; ref. 4.111.
E. angolensis   Angolan epauletted fruit   S, SW Angola, NW
                       bat   Namibia

Epauletted fruit bat
(Epomophorus wahlberg)

E. gambianus (*crypturus*) (*pousarguesi*) (*reii*)	Gambian epauletted fruit bat	Senegal – S Ethiopia – Angola, South Africa
E. grandis	Sanborn's epauletted fruit bat	NE Angola; Congo Rep.; ref. 4.111
E. labiatus (*anurus*)	Little epauletted fruit bat	Nigeria – Ethiopia – Malawi, Tanzania, ? Senegal
E. minor		S Sudan – S Somalia – Zambia, Malawi, Zanzibar
E. wahlbergi	Wahlberg's epauletted fruit bat	Cameroun – Somalia – S Africa, Angola, Zaire, Cameroun

Micropteropus; dwarf epauletted fruit bats; ref. 4.111.

M. intermedius	Hayman's epauletted fruit bat	NE Angola, SE Zaire
M. pusillus	Dwarf epauletted fruit bat	Senegal – Ethiopia, Angola – Zambia

Nanonycteris; ref. 4.111.

N. veldkampii	Veldkamp's dwarf fruit bat	Guinea – Congo Rep.

Scotonycteris

S. ophiodon	Pohle's fruit bat	Liberia – Congo Rep.
S. zenkeri	Zenker's fruit bat	Liberia – E Zaire

Casinycteris

C. argynnis	Short-palate fruit bat	Cameroun – E, NE Zaire

Cynopterus

C. archipelagus		Polillo I, off Luzon, Philippines (in *C. brachyotis* ?)
C. brachyotis (*minor*)	Lesser dog-faced fruit bat	S China, Burma – Java, Borneo, Philippines, Sulawesi, Sri Lanka, SW India, Andamans, Nicobars
C. horsfieldii	Horsfield's fruit bat	S Thailand – Java, Borneo
C. sphinx	Short-nosed fruit bat	? SE Pakistan; India – S China – Java, Andaman Is; Borneo
C. titthaecheilus (? *terminus*)		Sumatra, Krakatoa I, Nias I, Java; Bali; Lombok; ? Timor; ref. 4.12

Megaerops

M. ecaudatus	Tail-less fruit bat	Thailand – Sumatra, Borneo
M. kusnotoi		Java; K
M. niphanae		NE India, Thailand, Vietnam; refs. 4.12, 20
M. wetmorei		Mindanao, Philippines, N Borneo, Malaya; ref. 4.112

Ptenochirus

P. jagorii		Philippines
P. minor		Mindanao, Dinagat, Palawan Is, Philippines

Dyacopterus

D. brooksi		Sumatra, Luzon I, Philippines; K
D. spadiceus	Dyak fruit bat	Malaya – Borneo, ? Mindanao I, Philippines

Chironax

C. melanocephalus	Black-capped fruit bat	S Thailand – Java, Borneo, Sulawesi; refs. 4.14, 110

Latidens

L. salimalii		S India; K

Penthetor

P. lucasi	Lucas's short-nosed fruit bat	Malaya, Borneo

Thoopterus

T. nigrescens	Swift fruit bat	N Sulawesi, Sanghir Is, Morotai I, N Moluccas; ? Luzon

Aproteles

A. bulmerae		E New Guinea; K

Sphaerias

S. blanfordi	Blanford's fruit bat	NE India, S Tibet – NW Thailand, SW China

Balionycteris

B. maculata	Spotted-winged fruit bat	S Thailand, Malaya, Borneo

Aethalops

A. alecto	Pygmy fruit bat	Malaya, Sumatra, Java, Borneo; ref. 4.12

Haplonycteris
H. fischeri   Philippines

Alionycteris
A. paucidentata   Mindanao, Philippines; K

Otopteropus
O. cartilagonodus   Luzon, Philippines; K

Subfamily Harpyionycterinae

Harpyionycteris
H. celebensis   Sulawesi
H. whiteheadi   Harpy fruit bat   Philippines

Subfamily Nyctimeninae

Nyctimene; tube-nosed fruit bats; New Guinea, etc.
N. aello	Broad-striped tube-nosed bat	New Guinea
N. albiventer	Common tube-nosed bat	N Moluccas, New Guinea, Admiralty Is – Solomon Is, (?) NE Australia
N. celaeno		W, NW New Guinea; ref. 4.12
N. cephalotes	Pallas's tube-nosed bat	Sulawesi, Timor – W New Guinea, Admiralty Is
N. cyclotis	Round-eared tube-nosed bat	New Guinea; E New Britain
N. draconilla		S New Guinea; ref. 4.12
N. major	Greater tube-nosed bat	Bismarck Arch., Trobriand, Louisiade, D'Entrecasteaux, Solomon Is
N. malaitensis	Malaita tube-nosed bat	Malaita I, E Solomon Is
N. masalai		New Ireland; ref. 4.21
N. minutus	Lesser tube-nosed bat	Sulawesi, Buru, Obi I
N. rabori		Negros I, Philippines; ref. 4.22; E
N. robinsoni	Queensland tube-nosed bat	NE Australia
N. sanctacrucis		Santa Cruz Is
N. vizcaccia		Ruk I, Bismarck Arch; Solomon Is; ref. 4.21

Paranyctimene
P. raptor   Lesser tube-nosed bat   New Guinea

Subfamily Macroglossinae

Eonycteris; dawn fruit bats; SE Asia.

E. major		Borneo; Mentawei Is
E. robusta		Philippines; (in *E. major*?)
E. spelaea	Dawn bat	Burma – Java, Sumba,
(*rosenbergii*)	(Cave fruit bat)	Sulawesi, Philippines,
		Timor; refs. 4.12, 24, 110

Megaloglossus

M. woermanni	African long-tongued fruit bat	Guinea – Uganda – N Angola

Macroglossus; long-tongued fruit bats; SE Asia; ref. 4.12.

M. fructivorus		Mindanao, Philippines; (in *M. minimus*?)
M. minimus (*lagochilus*)	Common long-tongued fruit bat	Thailand, Indochina, Philippines – Solomons, NW, N Australia
M. sobrinus	Hill long-tongued fruit bat	NE India – Java, Bali

African long-tongued bat (Megaloglossus woermanni)

Syconycteris

S. australis (*naias*) (*crassa*)	Common blossom bat (Southern blossom bat)	S Moluccas, New Guinea, Bismarck Arch., Trobriand I, D'Entrecasteaux Is, NE Australia; ref. 4.12
S. carolinae		Halmahera, N Moluccas; ref. 4.23
S. hobbit		New Guinea; ref. 4.25

Melonycteris; (*Nesonycteris*).

Subgenus *Melonycteris*

M. melanops	Black-bellied fruit bat	Bismarck Arch., New Guinea?

Subgenus *Nesonycteris*

M. aurantius	Orange fruit bat	Florida, Choiseul Is, Solomons
M. woodfordi	Woodford's fruit bat	Solomons

Notopteris

N. macdonaldii	Long-tailed fruit bat	Vanuatu, New Caledonia, Fiji; ? Caroline Is

SUBORDER MICROCHIROPTERA

Family Rhinopomatidae

Mouse-tailed bats (rat-tailed bats, long-tailed bats); 3 species; Morocco, Senegal – Thailand, Sumatra; mainly desert and steppe; insectivorous.

Rhinopoma

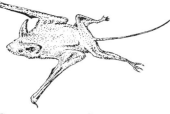

*Greater mouse-tailed bat
(Rhinopoma microphyllum)*

R. hardwickii	Lesser mouse-tailed bat	Morocco, Mauretania, Nigeria – Kenya – Thailand
R. microphyllum	Greater mouse-tailed bat	Senegal – India; Sumatra
R. muscatellum		S Arabia – W Pakistan

Family Emballonuridae

Sheath-tailed bats (sac-winged bats, pouched bats, ghost bats); *c.* 49 species; tropics and subtropics of world; insectivorous.

Subfamily Emballonurinae

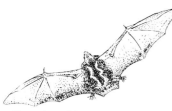

*Sac-winged bat
(Saccopteryx bilineata)*

Emballonura; Old-world sheath-tailed bats; Madagascar, S Burma – Pacific Islands.

E. alecto (*rivalis*)	Philippine sheath-tailed bat	Philippines, Borneo – S Moluccas, Tanimbar Is; ref. 4.143
E. atrata	Peters' sheath-tailed bat	Madagascar
E. beccarii	Beccari's sheath-tailed bat	New Guinea, etc.
E. dianae	Rennell Island sheath-tailed bat	New Guinea, New Ireland, Malaita, Rennell Is, Solomons; ref. 4.26
E. furax	Greater sheath-tailed bat	New Guinea, New Ireland; ref. 4.26; K
E. monticola	Lesser sheath-tailed bat	S Burma – Sulawesi
E. nigrescens		Sulawesi – Solomons
E. raffrayana	Raffray's sheath-tailed bat	Ceram, NW New Guinea, Solomons, Santa Cruz Is; ref. 4.26; K
E. semicaudata (*rotensis*)		Vanuatu, Mariana, Palau Is, Samoa, Fiji
E. sulcata		Caroline Is

Coleura

C. afra	African sheath-tailed bat	Africa, Aden
C. seychellensis	Seychelles sheath-tailed bat	Seychelles; ? Zanzibar; E

Rhynchonycteris

R. naso	Long-nosed bat (Tufted bat)	S Mexico – Bolivia, Brazil; Trinidad

Saccopteryx

S. bilineata	Greater white-lined bat (Sac-winged bat)	W, E Mexico – Bolivia, Brazil; Trinidad
S. canescens		Colombia – Peru, Brazil
S. gymnura		Brazil; ? Venezuela
S. leptura	Lesser white-lined bat	W Mexico – Peru, Brazil; Trinidad

Cormura

| C. brevirostris | Wagner's sac-winged bat | Nicaragua – Peru, Brazil |

Peropteryx

| P. kappleri | Greater sac-winged bat (Greater dog-like bat) | S Mexico – E Peru, Surinam, S, E Brazil |
| P. macrotis (*trinitatis*) | Lesser sac-winged bat (Lesser dog-like bat) | S Mexico – Peru, Paraguay, Tobago, Grenada, Trinidad, etc. |

Peronymus

| P. leucopterus | | Venezuela, Guianas – E Peru, Brazil |

Centronycteris

| C. maximiliani | Shaggy-haired bat (Thomas's bat) | S Mexico – Ecuador, ? E Peru, Brazil, Guianas |

Balantiopteryx; least sac-winged bats.

B. infusca		Ecuador
B. io	Thomas's least sac-winged bat	S Mexico – Guatemala, Belize
B. plicata	Peters' bat	N Mexico – Costa Rica

Taphozous; pouched bats, tomb bats, sheath-tailed bats; Africa, S Asia, Australia.

Subgenus *Taphozous*

T. australis	Gould's pouched bat	SE New Guinea, NE Queensland
T. georgianus (? *troughtoni*)	Sharp-nosed pouched bat	W, N Australia (in *T. melanopogon* ?)
T. hildegardeae	Hildegarde's tomb bat	Kenya, NE Tanzania, Zanzibar
T. hilli	Hill's pouched bat	W, N Australia; ref. 4.27
T. kapalgensis	White-striped sheath-tailed bat	N Australia; ref. 4.28
T. longimanus	Long-winged tomb bat	India – Java, Flores
T. mauritianus	Mauritian tomb bat	Africa S of Sahara, Madagascar, Aldabra, Mauritius, Reunion, Assumption

T. melanopogon	Black-bearded tomb bat	India – Java, Lesser Sundas, Borneo, Philippines
T. perforatus	Egyptian tomb bat	Senegal – Somalia, Mozambique, Zimbabwe – India
T. solifer		? China
T. theobaldi	Theobald's tomb bat	India – Vietnam; Java

Subgenus *Saccolaimus*

T. flaviventris	Yellow-bellied pouched bat	N, NC, E Australia
T. mixtus	Troughton's pouched bat	S, E New Guinea, N Queensland
T. peli	Pel's pouched bat	Liberia – NE Angola, Kenya
T. pluto (*capito*)		Philippines; (in *T. saccolaimus*?)
T. saccolaimus (*nudicluniatus*)		India – Sulawesi, Timor, New Guinea, Solomon Is, N, NE Australia

Subgenus *Liponycteris*

T. hamiltoni	Hamilton's tomb bat	Sudan, NW Kenya, ? S Chad; ref. 4.29
T. nudiventris	Naked-rumped tomb bat	Senegal – Somalia, Tanzania, Israel, Arabia, Burma, Cape Verde Is

Subfamily Diclidurinae

Diclidurus; ghost bats (white bats).

D. albus		E Peru, Venezuela, Surinam, Brazil, (?) Trinidad
D. ingens		Colombia – Guyana, NW Brazil
D. scutatus		Venezuela – Surinam, Peru, Brazil
D. virgo	White bat (Northern ghost bat)	W Mexico – Ecuador, Venezuela, Trinidad, Tobago

Depanycteris; (*Drepanycteris*); possibly a subgenus of *Diclidurus*.

D. isabella		Venezuela, Brazil

Cyttarops

C. alecto	Short-eared bat	Costa Rica, Nicaragua, Guyana, Brazil

Family Craseonycteridae

One species.

Craseonycteris

C. thonglongyai	Hog-nosed bat (Butterfly bat)	S Thailand; R

*Hog-nosed bat
(Craseonyctenis
thonglongyai)*

Family Nycteridae

Slit-faced bats, hollow-faced bats, hispid bats; *c.* 14 species; Africa, SW, SE Asia
to Java; chiefly forest and drier areas with trees; insectivorous, (carnivorous).

Nycteris; ref. 4.114

N. arge	Bate's slit-faced bat	Sierra Leone – SW Sudan, W Kenya, W Tanzania, NW Angola, Bioco
N. gambiensis	Gambian slit-faced bat	Senegal – Ghana, Togo (in *N. thebaica* ?)
N. grandis	Large slit-faced bat	Senegal – Kenya, Tanzania, Mozambique, Zimbabwe; ref. 4.29
N. hispida	Hairy slit-faced bat	Senegal – Ethiopia – S Africa
N. intermedia		Liberia – Zaire, Tanzania
N. javanica	Javan slit-faced bat	Java, Bali, Kangean I, (?) Timor
N. macrotis	Dobson's slit-faced bat (Greater slit-faced bat)	Senegal – Somalia – Zimbabwe, Mozambique
N. madagascariensis		Madagascar
N. major	Ja slit-faced bat	Benin – E, S Zaire, Gabon
N. nana	Dwarf slit-faced bat	Ghana – W Kenya – NE Angola
N. thebaica	Egyptian slit-faced bat	Africa S of Sahara, Morocco, Libya, Egypt, Israel, Jordan, Arabia, S Oman; Corfu (probably vagrant)
N. tragata		S Burma – Sumatra, Borneo
N. vinsoni	Vinson's slit-faced bat	Mozambique
N. woodi (*parisii*)	Wood's slit-faced bat	Cameroun; Somalia – Tanzania, Zambia, Zimbabwe, Transvaal

*Egyptian slit-faced bat
(Nycteris thebaica)*

*Yellow-winged bat
(Lavia frons)*

Family Megadermatidae

False vampire bats, yellow-winged bats; *c.* 5 species; Africa, SE Asia, Australia;
forest, savanna; insectivores, carnivores.

Megaderma; (*Lyroderma*).

M. lyra	Greater false vampire	E Afghanistan – S China, Malaya; Sri Lanka
M. spasma	Lesser false vampire	India – Java, Sulawesi, Philippines – Sangihe Is, N Moluccas, Sri Lanka; ref. 4.143

Macroderma

M. gigas	Australian false vampire (Ghost bat)	W, C, N Australia; (?) S New Guinea; V

Cardioderma

C. cor	Heart-nosed bat	Ethiopia – N Tanzania, Zanzibar

Lavia

L. frons	Yellow-winged bat	Senegal – Somalia – Zambia, Zanzibar

Family Rhinolophidae

Horseshoe bats; *c.* 64 species; Old-world tropics and temperate regions east to Australia; insectivores.

Rhinolophus; (*Rhinomegalophus*).

Lesser horseshoe bat (Rhinolophus hipposideros)

R. acuminatus	Acuminate horseshoe bat	Thailand – Java – Lombok, Borneo, Palawan
R. adami		Congo Republic
R. affinis	Intermediate horseshoe bat	N India – S China – Java, Lesser Sundas
R. alcyone	Halcyon horseshoe bat	Senegal – NE Zaire, Gabon
R. anderseni		Palawan, ? Luzon, Philippines
R. arcuatus (*toxopei*)	Arcuate horseshoe bat	Philippines, Borneo, Sumatra, Sulawesi, Moluccas, New Guinea
R. blasii	Blasius' horseshoe bat	N Africa, Italy – Afghanistan, Ethiopia – Transvaal, Natal, Mozambique
R. bocharicus		Turkestan, NE Iran, N Afghanistan
R. borneensis (*importunus*)	Bornean horseshoe bat	Kampuchea, ? Vietnam, Borneo, Java; ref. 4.12
R. canuti		Java, Timor; ref. 4.115

R. capensis	Cape horseshoe bat	S Africa
R. celebensis (*javanicus*) (*madurensis*) (*parvus*)	Sulawesi horseshoe bat	Sulawesi, Sangihe Is, Talaud Is, Java, Kangean Is, Timor; ref. 4.12, 143
R. clivosus	Geoffroy's horseshoe bat	Arabia – Algeria, Ethiopia – Zambia, S Africa
R. coelophyllus	Croslet horseshoe bat (Peters' horseshoe bat)	Burma – Malaya
R. cognatus		Andaman Is
R. cornutus	Little Japanese horseshoe bat	Japan, Ryukyu Is; ? E China
R. creaghi	Creagh's horseshoe bat	Borneo, Madura I
R. darlingi	Darling's horseshoe bat	Nigeria, Tanzania – Namibia – Mozambique
R. deckenii	Decken's horseshoe bat	Kenya, Tanzania
R. denti	Dent's horseshoe bat	Guinea; Mozambique – Namibia, S Africa
R. eloquens		S Sudan – S Somalia – N Tanzania
R. euryale	Mediterranean horseshoe bat	Portugal, Morocco – Iran
R. euryotis	Broad-eared horseshoe bat	Sulawesi, Moluccas, New Guinea, New Britain, Kiriwina I, Aru Is
R. ferrumequinum	Greater horseshoe bat	Britain, Morocco – N India, Japan
R. fumigatus	Rüppell's horseshoe bat	Senegal – N Ethiopia – S Africa
R. guineensis		Senegal – Sierra Leone; ref. 4.30
R. hildebrandtii	Hildebrandt's horseshoe bat	Ethiopia, Somalia – Botswana, Transvaal
R. hipposideros	Lesser horseshoe bat	Britain, Ireland – N India, NE Africa
R. imaizumii		Iriomote I, Ryukyu Is; ref. 4.31
R. inops		Mindanao, Philippines
R. keyensis	Insular horseshoe bat	Moluccas, Wetter I, Kei Is
R. landeri	Lander's horseshoe bat	Senegal – Somalia – Angola, Transvaal, Mozambique
R. lepidus (*refulgens*) (*feae*) (*osgoodi*)	Blyth's horseshoe bat	Afghanistan – S China – Sumatra; ref. 4.31
R. luctus	Woolly horseshoe bat	India, Sri Lanka – S China – Java, Bali, Borneo

R. maclaudi (*hilli*) (*ruwenzorii*)	Maclaud's horseshoe bat	Guinea, E Zaire, W Uganda, Rwanda
R. macrotis (*hirsutus*)	Big-eared horseshoe bat	Nepal – Malaya, Sumatra, Philippines
R. malayanus	N Malayan horseshoe bat	Thailand – Vietnam, Malaya
R. marshalli	Marshall's horseshoe bat	Thailand, Vietnam, N Malaya; ref. 4.147
R. megaphyllus	Southern horseshoe bat (Eastern horseshoe bat)	New Guinea, New Ireland, New Britain, E Australia
R. mehelyi	Mehely's horseshoe bat	Portugal, Morocco – Afghanistan
R. mitratus		C India; status doubtful
R. monoceros		Taiwan
R. nereis		Anamba Is, Natuna Is
R. paradoxolophus	Bourret's horseshoe bat	Thailand, Vietnam
R. pearsonii	Pearson's horseshoe bat	NE India – S China, Indochina; ref. 4.116
R. philippinensis	Philippines horseshoe bat (Large-eared horseshoe bat)	Philippines, Borneo – Sulawesi, Timor, NE Queensland
R. pusillus (*gracilis*) (*blythi*) (*minutillus*)	Least horseshoe bat	Tibet, NE India, S China – Java, Borneo
R. rex		S China
R. robinsoni	Peninsular horseshoe bat	NW Thailand, Malaya
R. rouxii (*petersi*)		India, S China, Vietnam, Sri Lanka
R. rufus		Philippines
R. sedulus	Lesser woolly horseshoe bat	Malaya, Borneo
R. shameli	Shamel's horseshoe bat	Burma – Cambodia
R. silvestris		Gabon, Congo Rep.
R. simplex	Lombok horseshoe bat	Lesser Sunda Is
R. simulator	Bushveld horseshoe bat	Nigeria, Cameroun – Ethiopia – Transvaal, etc.
R. stheno	Lesser brown horseshoe bat	Thailand – Java
R. subbadius		N India – Vietnam
R. subrufus		Philippines
R. swinnyi	Swinny's horseshoe bat	S Zaire, Zanzibar – S Africa
R. thomasi	Thomas's horseshoe bat	Burma, Yunnan – Indochina
R. trifoliatus	Trefoil horseshoe bat	N India – Java, Borneo
R. virgo		S Philippines
R. yunanensis	Dobson's horseshoe bat	NE India, Yunnan, Thailand; ref. 4.116

Family Hipposideridae

Old-world leaf-nosed bats, trident bats; *c.* 66 species; tropics and subtropics of Africa, S Asia – Philippines, Vanuatu, N Australia; forest, savanna; insectivores.

Diadem leaf-nosed bat (Hipposideros diadema)

Hipposideros; (*Syndesmotis*)

H. abae	Aba leaf-nosed bat	Guinea – S Sudan, Uganda
H. armiger	Himalayan leaf-nosed bat	N India – Malaya, Taiwan
H. ater	Dusky leaf-nosed bat	India – Java – Bismarck Arch., NW, N Australia; Philippines
H. beatus	Dwarf leaf-nosed bat	Liberia – SW Sudan, Gabon
H. bicolor	Bicolored leaf-nosed bat	S Thailand – Java, Lesser Sunda Is, Philippines; refs. 4.118, 143
H. breviceps		N Pagi I, Mentawei Is
H. caffer	Sundevall's leaf-nosed bat	Africa, Arabia
H. calcaratus (*cupidus*)	Spurred leaf-nosed bat	New Guinea, Bismarck Arch. – Solomon Is; ref. 4.32
H. camerunensis	Greater cyclops bat	Cameroun, E Zaire, Kenya
H. cervinus	Gould's leaf-nosed bat	Malaya, Sumatra, Philippines – Vanuatu, NE Australia; ref. 4.33
H. cineraceus (? *wrighti*)	Least leaf-nosed bat	N India – Malaya, Borneo, ? Philippines
H. commersoni	Commerson's leaf-nosed bat	Gambia – Somalia – Angola, Mozambique; Transvaal, Madagascar
H. coronatus		Philippines
H. corynophyllus		E New Guinea, ref. 4.117
H. coxi	Cox's leaf-nosed bat	Sarawak, Borneo
H. crumeniferus	Timor leaf-nosed bat	Timor I; status doubtful
H. curtus	Short-tailed leaf-nosed bat	Cameroun, Nigeria, Bioco
H. cyclops	Cyclops leaf-nosed bat	Guinea – Gabon – SW Kenya; Bioco
H. diadema	Diadem leaf-nosed bat	Burma – Java, Philippines, Borneo, New Guinea, Solomons, N Australia
H. dinops	Fierce leaf-nosed bat	Sulawesi, Peling I, Solomons
H. doriae		Sarawak, Borneo
H. durgadasi		C India; refs. 4.34, 35
H. dyacorum	Dyak leaf-nosed bat	Malaya, Borneo; ref. 4.146
H. fuliginosus	Sooty leaf-nosed bat	Guinea – Cameroun, Ethiopia, Zaire
H. fulvus	Fulvous leaf-nosed bat	Afghanistan – India – Sri Lanka
H. galeritus	Cantor's leaf-nosed bat (Fawn leaf-nosed bat)	Sri Lanka, India – Java, Borneo; ref. 4.33

H. halophyllus		Thailand; ref. 4.36
H. inexpectatus	Crested leaf-nosed bat	N Sulawesi
H. jonesi	Jones' leaf-nosed bat	Guinea – Nigeria
H. lamottei		Mt Nimba, Guinea; ref. 4.37
H. lankadiva		India, Sri Lanka
H. larvatus	Horsfield's leaf-nosed bat	N India, S China – Java, Sumba, Borneo
H. lekaguli	Lekagul's leaf-nosed bat	S Thailand, Malaya
H. lylei	Shield-faced leaf-nosed bat	Burma – Malaya
H. macrobullatus		Sulawesi; Kangean Is, Seram; ref. 4.118
H. maggietaylorae	Maggie's leaf-nosed bat	New Guinea, Bismarck Arch.; ref. 4.32
H. marisae	Aellen's leaf-nosed bat	Ivory Coast – Guinea
H. megalotis	Big-eared leaf-nosed bat	Ethiopia, Kenya
H. muscinus	Fly River leaf-nosed bat	Papua New Guinea
H. nequam	Malay leaf-nosed bat	Malaya
H. obscurus		Philippines
H. papua	Geelvink Bay leaf-nosed bat	N Moluccas, Biak I, W New Guinea; ref. 4.143; R
H. pomona		India – Thailand, S China; ref. 4.118
H. pratti	Pratt's leaf-nosed bat	SW China, Vietnam
H. pygmaeus		Philippines
H. ridleyi	Ridley's leaf-nosed bat	Malaya, Borneo; ref. 4.12; I
H. ruber	Noack's leaf-nosed bat	Senegal – Ethiopia – Angola
H. sabanus	Sabah leaf-nosed bat	Malaya, Sumatra, Borneo
H. schistaceus		India
H. semoni	Semon's leaf-nosed bat	E New Guinea, NE Queensland
H. speoris	Schneider's leaf-nosed bat	India, Sri Lanka
H. stenotis	Narrow-eared leaf-nosed bat	NW, N Australia
H. turpis	Lesser leaf-nosed bat	S Thailand, Ryukyu Is
H. wollastoni	Wollaston's leaf-nosed bat	New Guinea
Asellia		
A. patrizii	Patrizi's trident bat	Ethiopia
A. tridens	Trident bat	Morocco, Senegal – Pakistan
Aselliscus		
A. stoliczkanus	Stoliczka's trident bat	Burma, S China – Indochina, Penang I
A. tricuspidatus	Temminck's trident bat	Moluccas – Vanuatu

Anthops

| A. ornatus | Flower-faced bat | Solomon Is |

Cloeotis; ref. 4.38.

| C. percivali | Short-eared trident bat (Percival's trident bat) | Kenya – SE Botswana – Mozambique, Transvaal, Swaziland |

Rhinonycteris; (*Rhinonicteris*); ref. 4.38.

| R. aurantius | Orange leaf-nosed bat | NW, N Australia; K |

Triaenops; ref. 4.38.

| T. furculus | Trouessart's trident bat | Madagascar, Aldabra, Picard, Cosmoledo Is |
| T. persicus (? humbloti) (? rufus) | Persian trident bat | Congo Rep. – Angola, Mozambique – S Arabia – Pakistan; Madagascar |

Coelops

| C. frithii | Tail-less leaf-nosed bat | E India, S China – Java, Bali, Taiwan |
| C. robinsoni (hirsutus) | Malayan tail-less leaf-nosed bat | S Thailand, Malaya, Borneo, Mindoro, Philippines |

Paracoelops

| P. megalotis | | Vietnam |

Family Noctilionidae

Bulldog bats (hare-lipped bats, mastiff bats, fisherman bats); 2 species; New-world tropics; insectivorous, piscivorous.

Noctilio

| N. albiventris | Lesser bulldog bat | Honduras – N Argentina |
| N. leporinus | Greater bulldog bat (Fisherman bat) | SW, S Mexico – N Argentina, Antilles, Trinidad, S Bahamas |

Fisherman bat
(Noctilio leporinus)

Family Mormoopidae

Naked-backed bats, moustached bats, ghost-faced bats; *c.* 8 species; S USA – Brazil, Greater and Lesser Antilles; insectivorous.

Pteronotus

Subgenus *Pteronotus*

| P. davyi | Davy's naked-backed bat | Mexico – Peru, Lesser Antilles, Trinidad |
| P. gymnonotus (suapurensis) | Big naked-backed bat | S Mexico – E Peru, Brazil |

Davy's naked-backed bat
(Pteronotus davyi)

Subgenus *Chilonycteris*

P. macleayii	Macleay's moustached bat	Cuba, I of Pines, Jamaica
P. personatus	Wagner's moustached bat	N Mexico – Brazil, Trinidad
P. quadridens (*fuliginosus*)	Sooty moustached bat	Greater Antilles; ref. 4.39

Subgenus *Phyllodia*

P. parnellii	Parnell's moustached bat	NE Mexico – E Peru, Brazil, Trinidad, Greater Antilles

Mormoops; (*Aello*).

M. blainvillii (*cuvieri*)	Antillean ghost-faced bat	Greater Antilles; † C Bahamas; ref. 4.40
M. megalophylla	Peters' ghost-faced bat	S USA – Peru, Venezuela, Trinidad, † Cuba

California leaf-nosed bat (Macrotus californicus)

Family Phyllostomidae

New-world leaf-nosed bats; *c.* 152 species; New-world tropics and subtropics; insectivorous, carnivorous, frugivorous; mainly forest.

Subfamily Phyllostominae

Micronycteris; big-eared bats; Mexico – Peru, Brazil.

Subgenus *Micronycteris*

M. megalotis	Brazilian big-eared bat	W, S Mexico – E Venezuela, Brazil, Peru; Trinidad, etc., Grenada
M. minuta	Gervais' big-eared bat	Nicaragua – E Peru, Bolivia, Brazil, Trinidad
M. schmidtorum	Schmidt's big-eared bat	S Mexico, Guatemala – Colombia

Subgenus *Xenoctenes*; (possibly synonymous with *Micronycteris*); ref. 4.41.

M. hirsuta	Hairy big-eared bat	Honduras – E Peru, Brazil, Trinidad

Subgenus *Lampronycteris*

M. brachyotis	Dobson's big-eared bat	S Mexico – C Brazil; Trinidad

Subgenus *Neonycteris*

M. pusilla		E Colombia, NW Brazil

Subgenus *Trinycteris*

| *M. nicefori* | Niceforo's big-eared bat | S Mexico – NE Peru, Brazil, Trinidad |

Subgenus *Glyphonycteris*

| *M. behnii* | | S Peru, C Brazil |
| *M. sylvestris* | Brown big-eared bat (Large-eared forest bat) | W, S Mexico – E Peru, SE Brazil, Trinidad |

Barticonycteris; (*Micronycteris*).

| *B. daviesi* | Davies' large-eared bat | Costa Rica, Panama, E Peru, Guyana, Surinam, Brazil; refs. 4.42, 43 |

Macrotus; big-eared bats; leaf-nosed bats; SW USA – Guatemala, W Indies.

| *M. californicus* | California leaf-nosed bat | S California, S Nevada, Arizona, NW Mexico; (in *M. waterhousii*?); ref. 4.40 |
| *M. waterhousii* | Waterhouse's leaf-nosed bat | N Mexico – Guatemala, Bahamas, Greater Antilles |

Lonchorhina; sword-nosed bats; S Mexico – Peru, Brazil, Trinidad.

L. aurita	Tomes' long-eared bat	S Mexico – E Peru, Bolivia, Brazil, Trinidad, Bahamas (accidental; ref. 4.7)
L. fernandezi		Venezuela; ref. 4.44
L. marinkellei		E Colombia
L. orinocensis		C Venezuela, Colombia

Macrophyllum

| *M. macrophyllum* | Long-legged bat | S Mexico – N Argentina |

Tonatia; round-eared bats; S Mexico – N Argentina.

T. bidens	Spix's round-eared bat	Guatemala – E Peru, E Brazil, Trinidad, Jamaica
T. brasiliense		Venezuela – Brazil; Trinidad
T. carrikeri		Venezuela, E Peru, Bolivia, Guianas
T. evotis	Davis's round-eared bat	S Mexico – Honduras
T. minuta (nicaraguae)	Pygmy round-eared bat	S Mexico – E Peru; (in *T. brasiliense*?); ref. 4.45
T. schulzi		Guianas, N Brazil; ref. 4.46
T. silvicola	D'Orbigny's round-eared bat	Honduras – N Argentina

T. venezuelae		Venezuela; (in *T. brasiliense*?); ref. 4.45

Mimon; Gray's spear-nosed bats; S Mexico – Bolivia, Brazil.

Subgenus *Mimon*

M. bennettii	Bennett's spear-nosed bat	Guyana, Surinam, SE Brazil
M. cozumelae	Cozumel spear-nosed bat	S Mexico – N Colombia; (in *M. bennettii*?)

Subgenus *Anthorhina*

M. crenulatum (*koepckeae*)	Striped spear-nosed bat	S Mexico – Peru, Bolivia, Brazil, Trinidad

Phyllostomus; spear-nosed bats; S Mexico – Bolivia, N Argentina.

P. discolor	Pale spear-nosed bat	S Mexico – N Argentina
P. elongatus		Colombia, Guianas – E Peru, Bolivia, SE Brazil
P. hastatus	Greater spear-nosed bat	Honduras – Peru – Paraguay; Trinidad
P. latifolius		SE Colombia, Guyana, Surinam, C Brazil

Phylloderma; perhaps in *Phyllostomus*; ref. 4.148.

P. stenops	Peters' spear-nosed bat	S Mexico – Peru, NE Brazil; Surinam, Trinidad

Trachops

T. cirrhosus	Fringe-lipped bat	S Mexico – E Peru, Bolivia, S Brazil; Trinidad

Chrotopterus

C. auritus	Peters' woolly false vampire bat	S Mexico – Paraguay, N Argentina

Vampyrum

V. spectrum	American false vampire bat (Linnaeus's false vampire bat)	S Mexico – Peru, SW Brazil, Trinidad, Jamaica (accidental ?)

Subfamily Glossophaginae

Glossophaga; long-tongued bats; Mexico – N Argentina, W Indies, Bahamas.

G. commissarisi	Commissari's long-tongued bat	W Mexico – Panama; Colombia, N Peru, W Brazil

G. leachii (*alticola*)	Davis's long-tongued bat	C Mexico – Costa Rica; ref. 4.47
G. longirostris	Miller's long-tongued bat	Colombia, Venezuela, Lesser Antilles, Trinidad, etc.; ref. 4.119
G. morenoi (*mexicana*) (*brevirostris*)		S Mexico; refs. 4.47, 120
G. soricina	Pallas's long-tongued bat	N Mexico – N Argentina, Trinidad etc., Jamaica, Bahamas

Monophyllus

M. plethodon	Barbados long-tongued bat (Insular long-tongued bat)	Lesser Antilles, † Puerto Rico
M. redmani	Jamaican long-tongued bat (Leach's long-tongued bat)	Cuba, Hispaniola, Puerto Rico, Jamaica, S Bahamas

Leptonycteris; Saussure's long-tongued bats; SW USA – Venezuela; ref. 4.149.

L. curasoae		Colombia, Venezuela, Curaçao I, etc.
L. nivalis	Big long-nosed bat	SE Arizona, W Texas – Mexico, ? Guatemala; V
L. yerbabuenae (*sanborni*)	Little long-nosed bat	S Arizona, New Mexico – El Salvador; in *L.* *curasoae* ?; ref. 4.149; V

Big long nosed bat
(Leptonycteris nivalis)

Lonchophylla

L. bokermanni		SE Brazil
L. concava	Goldman's long-tongued bat	Costa Rica, Panama, ? Peru; (in *L. mordax*?)
L. dekeyseri		Brazil; ref. 4.48
L. handleyi		Colombia, Ecuador, Peru; ref. 4.49
L. hesperia		Peru, Ecuador
L. mordax	Brazilian long-tongued bat	Colombia, Ecuador, Bolivia, Brazil, ? Peru
L. robusta	Panama long-tongued bat	Nicaragua – Peru
L. thomasi	Thomas's long-tongued bat	Panama, Peru, Bolivia, Brazil, Guianas

Lionycteris

L. spurrelli	Little long-tongued bat	Panama, E Peru, Ecuador, N Brazil, Venezuela, Guianas

Anoura; Geoffroy's long-nosed bats; Mexico – Brazil, NW Argentina.

A. caudifera		Colombia, Venezuela – E Peru, E Brazil, N Argentina
A. cultrata (*brevirostrum*) (*werckleae*)		Costa Rica, Panama, Venezuela, Colombia, Peru
A. geoffroyi	Geoffroy's tail-less bat	N Mexico – NW Argentina; Trinidad, Grenada
A. latidens		Venezuela – Peru; ref. 4.50

Scleronycteris

S. ega		Venezuela, Brazil

Lichonycteris

L. obscura (*degener*)	Brown long-nosed bat	Guatemala – Peru, Bolivia, Surinam; ref. 4.121

Hylonycteris

H. underwoodi	Underwood's long-tongued bat	W, S Mexico – W Panama

Platalina

P. genovensium		Peru

Choeroniscus; long-tailed bats; Mexico – Peru, Brazil.

C. godmani	Godman's bat	W Mexico – Colombia – Surinam
C. intermedius		Trinidad, E Peru, Brazil, Guianas
C. minor (*inca*)		Bolivia – Colombia – Guianas, Brazil
C. periosus		W Colombia

Choeronycteris

C. mexicana	Mexican long-tongued bat	S California, S Arizona, S New Mexico – Honduras, (?) NW Venezuela

Musonycteris; (*Choeronycteris*).

M. harrisoni	Banana bat	W, C Mexico

Subfamily Carolliinae

Carollia; short-tailed leaf-nosed bats; Mexico – E Peru, Bolivia, S Brazil.

C. brevicauda	Silky short-tailed bat	E Mexico – E Peru, Bolivia, E Brazil
C. castanea	Allen's short-tailed bat	Honduras – E Peru, Bolivia, Venezuela, Guianas

C. perspicillata	Seba's short-tailed bat	S Mexico – S Brazil, Paraguay; Trinidad, Tobago, Grenada, ? Jamaica, (?) N Argentina
C. subrufa	Hahn's short-tailed bat	W Mexico – Nicaragua

Rhinophylla

R. alethina		W Colombia, Ecuador
R. fischerae		Colombia – E Peru, NW Brazil
R. pumilio		Bolivia – Colombia – Guianas

(Rhinophylla pumilio)

Subfamily Sturnirinae

Sturnira; yellow-shouldered bats, American epauletted bats; Mexico – N Argentina, Uruguay, ? Chile, Trinidad, Lesser Antilles, Jamaica.

Subgenus *Sturnira*

S. aratathomasi		Venezuela, W Ecuador
S. bogotensis		Bolivia – Venezuela
S. erythromos		Bolivia – Venezuela
S. lilium	Yellow-shouldered bat	N Mexico – N Argentina, Uruguay, ? Chile; Trinidad, Lesser Antilles, ? Jamaica
S. ludovici	Anthony's bat	N Mexico – Ecuador, Venezuela
S. luisi		Costa Rica – Peru; ref. 4.51
S. magna		Amazonian Colombia – Bolivia
S. mordax	Hairy-footed bat (Talamancan bat)	Costa Rica
S. oporaphilum		Bolivia; ref. 4.52
S. thomasi	Sofaian bat (Thomas's epauletted bat)	Guadeloupe, Lesser Antilles
S. tildae		Bolivia – Guianas, Brazil, Trinidad

Subgenus *Corvira*

S. bidens		E Peru – Venezuela, Brazil
S. nana		S Peru

Subfamily Stenoderminae

Uroderma; tent-building bats; Mexico – S Peru, Bolivia, SE Brazil, Trinidad.

U. bilobatum	Tent-building bat (Tent-making bat)	S Mexico – S Peru, Bolivia, SE Brazil, Trinidad

Tent-building bat (Uroderma bilobatum)

U. magnirostrum	Davis' bat	S Mexico – E Peru, N Bolivia, Brazil

Vampyrops; (*Platyrrhinus*); broad-nosed bats, white-lined bats; S Mexico – N Argentina, Paraguay, Uruguay.

V. aurarius		E Venezuela – Surinam; (in *V. dorsalis*?)
V. brachycephalus		Colombia – E Peru, Guianas
V. dorsalis	Thomas's broad-nosed bat	Panama – Bolivia, ? Venezuela
V. helleri	Heller's broad-nosed bat	S Mexico – Paraguay, Brazil, Trinidad
V. infuscus		Colombia – Bolivia, Brazil
V. lineatus (*recifinus*)		Colombia, E Peru, Guianas, C, E Brazil – Uruguay
V. umbratus (*oratus*) (*aquilus*)		Colombia, Venezuela; ref. 4.53
V. vittatus	Greater broad-nosed bat	Costa Rica – Bolivia, Venezuela

Vampyrodes

V. caraccioli	Great striped-faced bat	Venezuela – Surinam, Trinidad, Tobago
V. major (*ornatus*)	San Pablo bat	S Mexico – Bolivia, N Brazil; (in *V. caraccioli* ?); ref. 4.40

Vampyressa; yellow-eared bats; S Mexico – E Peru, Brazil, Guyana.

Subgenus *Vampyressa*

V. melissa		Panama, Colombia, Peru
V. pusilla	Little yellow-eared bat	S Mexico – Peru, SE Brazil, E Paraguay

Subgenus *Vampyriscus*

V. bidens		Colombia – Bolivia, N Brazil, Guyana, Surinam
V. brocki		Colombia – Surinam, Brazil
V. nymphaea	Big yellow-eared bat	Nicaragua – W Colombia

Chiroderma; white-lined bats; big-eyed bats; Mexico – Peru, Bolivia, Brazil, Trinidad, Guadeloupe, Lesser Antilles.

C. doriae		E Brazil
C. improvisum	Guadeloupe white-lined bat	Guadeloupe, Montserrat, Lesser Antilles

C. salvini	Salvin's white-lined bat	W, C Mexico – Bolivia, Venezuela
C. trinitatum	Goodwin's bat	Panama, E Peru, Bolivia, Brazil, Trinidad, Tobago
C. villosum	Shaggy-haired bat	S Mexico – E Peru, Bolivia, Brazil, Trinidad, Tobago

Ectophylla; refs. 4.54, 55.

E. alba	White bat	Nicaragua – W Panama

Mesophylla; refs. 4.55, 56.

M. macconnelli	McConnell's bat	Costa Rica – E Peru, Bolivia, Brazil, Trinidad

Artibeus; American fruit bats; Mexico – Peru, Bolivia, Brazil, Trinidad, Bahamas, Antilles; ref. 4.122.

*Jamaican fruit-eating bat
(Artibeus jamaicensis)*

A. amplus		Colombia, Venezuela, ? Guyana; ref. 4.122
A. anderseni		Bolivia – Peru, W Brazil, French Guiana
A. aztecus	Highland fruit-eating bat	W, C Mexico – W Panama
A. cinereus	Gervais' fruit-eating bat	Venezuela, Brazil, Guianas, Trinidad, Tobago, Grenada
A. concolor		Colombia, Venezuela – E Peru, Brazil, Guianas
A. fraterculus		S Ecuador, N Peru
A. fuliginosus		Peru – Colombia – Brazil, French Guiana
A. glaucus (*bogotensis*)		Venezuela – Peru, Bolivia, Guyana
A. gnomus		Venezuela, Peru, Brazil, Guyana; ref. 4.122
A. hirsutus	Hairy fruit-eating bat	W Mexico
A. inopinatus	Honduran fruit-eating bat	El Salvador, Honduras, Nicaragua; (in *A. hirsutus?*)
A. intermedius		C, NE Mexico – Panama, Colombia; ref. 4.57
A. jamaicensis	Jamaican fruit-eating bat	Mexico – Paraguay, Brazil, Bahamas, Antilles, Trinidad
A. lituratus	Great fruit-eating bat	S Mexico – N Argentina, Lesser Antilles, Trinidad, Tobago
A. phaeotis	Dwarf fruit-eating bat (Pygmy fruit-eating bat)	W Mexico – E Peru; Guyana
A. planirostris		Guianas – Colombia – N Argentina

A. pumilio		Peru
A. toltecus	Lowland fruit-eating bat	C Mexico – Panama
A. watsoni	Thomas's fruit-eating bat	S Mexico – Colombia

Enchisthenes; possibly a subgenus of *Artibeus*.

E. hartii	Little fruit-eating bat	NE Mexico – Bolivia; S Arizona, Trinidad

Ardops

A. nichollsi	Tree bat	Lesser Antilles

Phyllops; fig-eating bats, falcate-winged bats.

P. falcatus	Cuban fig-eating bat	Cuba; † I of Pines
P. haitiensis	Dominican fig-eating bat	Hispaniola

Ariteus

A. flavescens	Jamaican fig-eating bat	Jamaica

Stenoderma

S. rufum	Red fruit bat (Desmarest's fig-eating bat)	Puerto Rico, Virgin Is

Pygoderma

P. bilabiatum	Ipanema bat	Surinam – Paraguay, N Argentina

Ametrida

A. centurio		Venezuela, Guianas, Brazil, Trinidad, etc.

Sphaeronycteris

S. toxophyllum		Colombia, E Peru – Venezuela, Bolivia

Centurio

C. senex	Wrinkle-faced bat	W, NE Mexico – Venezuela, Trinidad, etc.

Subfamily Phyllonycterinae

Brachyphylla; ref. 4.58.

B. cavernarum	St Vincent fruit-eating bat (Antillean fruit-eating bat)	Puerto Rico, Lesser Antilles, Virgin Is
B. nana (*pumila*)	Cuban fruit-eating bat	Cuba, I of Pines, Grand Cayman, S Bahamas, Hispaniola, † Jamaica

Erophylla; Greater Antilles.

E. bombifrons	Brown flower bat	Hispaniola, Puerto Rico; refs. 4.40, 59

E. sezekorni	Buffy flower bat	Bahamas, Cuba, I of Pines, Jamaica, Cayman Is

Phyllonycteris

Subgenus *Phyllonycteris*

P. major†	Puerto Rican flower bat	Puerto Rico, probably extinct
P. poeyi (*obtusa*)	Cuban flower bat	Cuba, I of Pines, Hispaniola; K

Subgenus *Reithronycteris*

P. aphylla	Jamaican flower bat	Jamaica; K

Subfamily Desmodontinae

Vampire bats; S USA – C Chile, N Argentina, Uruguay.

Desmodus

D. rotundus	Common vampire bat	N Mexico – C Chile, N Argentina, Uruguay, Trinidad, † Cuba

Diaemus; possibly congeneric with *Desmodus*; ref. 4.56.

D. youngi	White-winged vampire	NE Mexico – E Peru – N Argentina, Brazil; Trinidad

Diphylla

D. ecaudata	Hairy-legged vampire bat	S Texas – E Peru, S Brazil

Family Natalidae

Funnel-eared bats, long-legged bats; *c.* 5 species; Mexico – Brazil, Bahamas, Antilles.

Natalus

Subgenus *Natalus*

N. stramineus (*major*) (*espiritosantensis*)	Mexican funnel-eared bat	N Mexico – Brazil, Guianas, Antilles, † Cuba; refs. 4.60, 61
N. tumidirostris		Colombia – Surinam; Trinidad, Curacao I

Subgenus *Chilonatalus*

N. micropus (*macer*) (*brevimanus*)	Cuban funnel-eared bat (Jamaican long-legged bat)	Jamaica, Cuba, Hispaniola, Old Providence I, off Nicaragua

Mexican funnel-eared bat (Natalus stramineus)

N. tumidifrons		Bahamas
Subgenus *Nyctiellus*		
N. lepidus	Gervais' funnel-eared bat	Bahamas, Cuba, Isle of Pines

Eastern smoky bat
(Furipterus horrens)

Spix's disk-winged bat
(Thyroptera tricolor)

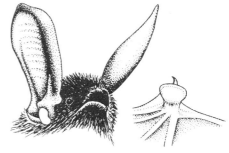

Sucker-footed bat
(Myzopoda aurita)

Family Furipteridae

Smoky bats, thumbless bats; 2 species; Panama – N Chile, Brazil, Guianas, Trinidad.

Furipterus

F. horrens	Eastern smoky bat	Costa Rica – E Peru, Guianas, SE Brazil, Trinidad

Amorphochilus

A. schnablii		W Ecuador – N Chile

Family Thyropteridae

Disc-winged bats, New-world sucker-footed bats; 2 species; S Mexico – Peru, S, E Brazil.

Thyroptera

T. discifera	Peters' disc-winged bat (Honduran disc-winged bat)	Nicaragua – Bolivia, Guianas, Brazil
T. tricolor	Spix's disc-winged bat	S Mexico – Bolivia, Guianas, S, E Brazil, Trinidad

Family Myzopodidae

One species.

Myzopoda

M. aurita	Sucker-footed bat	Madagascar; K

Family Vespertilionidae

Vespertilionid bats; *c*. 350 species; cosmopolitan except some small oceanic islands, Arctic and Antarctic regions beyond limits of tree growth.

Big brown bat
(*Eptesicus fuscus*)

Subfamily Vespertilioninae; refs. 4.123, 124

Myotis; mouse-eared bats, little brown bats; cosmopolitan, except some oceanic islands, Arctic, Antarctic; subgeneric classification unstable.

Subgenus *Myotis*

M. altarium		S China, N Thailand; ref. 4.126
M. blythii	Lesser mouse-eared bat	Spain, Morocco – N India – S China
M. bocagii	Rufous mouse-eared bat	Liberia – Kenya – Angola, Transvaal; S Yemen
M. chinensis	Large myotis	S China, N Thailand
M. emarginatus	Geoffroy's bat	SW Europe – Turkestan, Afghanistan, N Africa
M. goudoti	Malagasy mouse-eared bat	Madagascar; Anjouan I, Comoro Is.
M. morrisi		Ethiopia, Nigeria; ref. 4.151
M. myotis	Large mouse-eared bat	SW Europe – Asia Minor; Azores; ref. 4.67; K
M. pequinius		E China
M. sicarius		Nepal, Sikkim
M. tricolor (*loveni*)	Cape Hairy bat	Liberia, Ethiopia – S Africa; refs. 4.125, 1.144

Subgenus *Chrysopteron*

M. formosus (*rufopictus*) (*weberi*) (*bartelsii*)	Hodgson's bat (Orange-winged bat)	E Afghanistan – Korea, Taiwan, Philippines, Sulawesi, Java, Bali; (K)
M. hermani	Herman's bat	NW Sumatra; R
M. welwitschii	Welwitsch's hairy bat	Ethiopia – Angola, S Africa

Subgenus *Cistugo*

M. leseuri	Leseur's hairy bat	Cape Province, S Africa (in *M. seabrai* ?); ref. 4.66
M. seabrai	Angola wing-gland bat (Angola hairy bat)	Angola – Cape Province, S Africa

Subgenus *Selysius*

M. abei		Sakhalin I

M. aelleni		SW Argentina
M. annectans	Hairy-faced bat	NE India, N Thailand
M. atacamensis		S Peru, N Chile
M. ater		Siberut I, Borneo, Philippines, Sulawesi – New Guinea
M. australis		New South Wales (in *M. ater* ?); ref. 4.12
M. austroriparius	South-eastern myotis	N Carolina – Indiana – Louisiana, Florida
M. brandtii	Brandt's bat	France, Britain – E Siberia, Hokkaido; ref. 4.127
M. californicus	California myotis	Alaska – Guatemala
M. carteri		W Mexico; ref. 4.63
M. chiloensis		Chile
M. ciliolabrum		SW Canada – Nebraska, Oklahoma, W Mexico; ref. 4.107
M. cobanensis	Guatemalan myotis	Guatemala; status doubtful
M. dominicensis	Dominican myotis	Dominica, Lesser Antilles
M. elegans	Elegant myotis	Mexico – Costa Rica
M. findleyi	Findley's myotis	Tres Marias Is, W Mexico; ref. 4.63
M. fortidens	Cinnamon myotis	W Texas – Guatemala; ref. 4.65
M. frater		Turkestan – E Siberia, SE China, Japan
M. hosonoi		N, C Honshu, Japan
M. ikonnikovi		E Siberia, N Korea, Sakhalin, Hokkaido
M. insularum		? Samoa; status doubtful
M. keaysi	Hairy-legged myotis	NE Mexico – Venezuela – Peru; Trinidad
M. leibii	Least brown bat (Small-footed myotis)	SE Canada – Oklahoma, Georgia; ref. 4.107
M. levis		S Brazil – Paraguay, Uruguay, Argentina
M. martiniquensis	Schwarz's myotis	Martinique, Barbados, Lesser Antilles
M. montivagus	Burmese whiskered bat	S India – S China, Malaya, Borneo
M. muricola (*browni*) (*herrei*)		Afghanistan – Taiwan, Philippines, Moluccas, Lesser Sunda Is
M. mystacinus	Whiskered bat	Ireland – Japan, Tibet, N India; Morocco
M. nesopulos		NW Venezuela, Curacao I, Bonaire I; ref. 4.68

M. nigricans	Black myotis	W, NE Mexico – N Argentina, Paraguay; Trinidad, Tobago, Grenada
M. oreias	Singapore whiskered bat	Singapore I
M. oxyotus	Montane myotis	Costa Rica – Peru, N Bolivia
M. oxensis		C Honshu, Japan
M. patriciae		Mindanao, Philippines
M. peninsularis	Peninsular myotis	Baja California
M. planiceps	Flat-headed myotis	N Mexico
M. ridleyi	Ridley's bat	Malaya, Borneo, ? Sumatra
M. riparius	Riparian myotis	Honduras – Peru – Uruguay; Trinidad
M. rosseti	Thick-thumbed myotis	Thailand, Kampuchea
M. ruber		SE Brazil, Paraguay, N Argentina
M. scotti	Scott's mouse-eared bat	Ethiopia
M. siligorensis	Himalayan whiskered bat	N India – S China – Malaya, Borneo; ref. 4.69
M. sodalis	Social bat (Indiana myotis)	C E USA; V
M. surinamensis		? Surinam; status doubtful
M. velifer	Cave myotis	S USA – Honduras
M. volans	Long-legged myotis (Hairy-winged bat)	Alaska – S Mexico
M. yesoensis		Hokkaido, Japan; ref. 4.70
M. yumanensis	Yuma myotis	British Columbia – C USA – C Mexico

Subgenus *Hesperomyotis*

M. simus		Colombia – Paraguay

Subgenus *Paramyotis*

M. auriculus	Mexican long-eared myotis	New Mexico, Arizona – Guatemala
M. bechsteinii	Bechstein's bat	Spain, England – W Russia, Caucasus, Iran
M. evotis	Long-eared myotis	NW Mexico
M. keenii (*septentrionalis*)	Keen's myotis	Alaska – Washington, Manitoba – Newfoundland – Florida
M. milleri	Miller's myotis	Baja California

Subgenus *Isotus*

M. bombinus		E Asia, Japan; ref. 4.62
M. nattereri	Natterer's bat	W Europe – Urals – Israel; N Africa

Large mouse-eared bat
(Myotis myotis)

M. schaubi (*araxenus*)		Armenia, W Iran; ref. 4.62
M. thysanodes	Fringed myotis	SW Canada – S Mexico

Subgenus *Leuconoe*

M. adversus	Long-footed bat	Taiwan, Malaya – Solomon Is, Vanuatu, N, E Australia
M. albescens	Silver-tipped myotis	S Mexico – N Argentina, Uruguay
M. capaccinii	Long-fingered bat	Spain – Uzbekistan; N Africa; V
M. dasycneme	Pond bat	E France – Manchuria; K
M. daubentonii (*petax*) (*nathalinae*)	Daubenton's bat	Portugal, Britain – E Siberia, Manchuria, Sakhalin, Japan; ref. 4.64
M. fimbriatus		Fujian, SE China
M. grisescens	Grey myotis	Oklahoma – Kentucky – Georgia; E
M. hasseltii		Sri Lanka; Burma – Malaya, Java, Borneo
M. horsfieldii (*dryas*) (*peshwa*) (? *jeannei*)		India – S China, Java, Bali, Sulawesi, ? Philippine Is
M. longipes		Afghanistan, Kashmir, ? Vietnam
M. lucifugus	Little brown myotis	Alaska, SE Canada – C Mexico
M. macrodactylus		E Siberia, S Kurile Is, Japan
M. macrotarsus		Philippines, Borneo
M. pruinosus		N Honshu, Japan
M. stalkeri		Kei Is, New Guinea

Subgenus *Rickettia*

M. ricketti	Rickett's big-footed bat	E China

Pizonyx; (probable subgenus of *Myotis*)

P. vivesi	Fish-eating bat	NW Mexico; coasts, islands

Lasionycteris

L. noctivagans	Silver-haired bat	Alaska, S Canada — NE Mexico; Bermuda

Eudiscopus

E. denticulous	Disc-footed bat	Burma, Laos

Pipistrellus; (*Vespertilio*, *Perimyotis*, refs. 4.71, 123, 124); pipistrelles, serotines; Europe, Africa — Solomon Is, Australia, Tasmania; S Canada – Honduras.

(Pipistrellus subflavus)

Subgenus *Pipistrellus*

P. abramus		E Siberia – Vietnam, Taiwan; ref. 4.123
P. adamsi		N Australia; ref. 4.128
P. aero		NW Kenya, ? Ethiopia
P. angulatus	Greater New Guinea pipistrelle	New Guinea, Bismarck Arch., Solomons; refs. 4.12, 72, 128
P. babu		Afghanistan, Xizang, N, C India – Burma, ? China
P. ceylonicus	Kelaart's pipistrelle	Pakistan – S China, Vietnam, Borneo, Sri Lanka
P. collinus		Papua New Guinea; ref. 4.128
P. coromandra	Indian pipistrelle	E Afghanistan – S China, Sri Lanka, Nicobar Is
P. crassulus	Broad-headed pipistrelle	Cameroun, C Zaire
P. deserti	Desert pipistrelle	Algeria – Egypt, N Sudan, Burkina Faso
P. endoi		Honshu, Japan
P. inexspectatus	Aellen's pipistrelle	Benin – Uganda, ? Sudan; ref. 4.74
P. javanicus	Javan pipistrelle	Burma, Thailand – Java, Flores I, Lesser Sunda Is, Borneo, Sulawesi, Talaud Is, Philippines, ? N Australia; ref. 4.143
P. kuhlii	Kuhl's pipistrelle	SW Europe – Kashmir, Africa; Cape Verde Is
P. maderensis	Madeira pipistrelle	Madeira, Canary Is
P. mimus	Indian pygmy pipistrelle	Afghanistan – Thailand, S China; Sri Lanka
P. minahassae	Minahassa pipistrelle	N Sulawesi
P. murrayi		Christmas I, Indian Ocean, ? Cocos Keeling Is; (in *P. tenuis* ?)
P. nanulus	Tiny pipistrelle	Sierra Leone – Uganda; ref. 4.74
P. nathusii	Nathusius' pipistrelle	Spain – Urals, Caucasus
P. paterculus		NE India – S China; ref. 4.123
P. peguensis		Pegu, S Burma
P. permixtus	Dar-es-Salaam pipistrelle	Tanzania
P. pipistrellus	Common pipistrelle	W Europe, Morocco – Kashmir, ? Korea, Japan, Taiwan

P. rueppelli	Ruppell's bat	N Africa, Senegal – Tanzania – Transvaal, Egypt, Israel, Iraq
P. rusticus	Rusty bat	Ghana – Ethiopia – Namibia – Transvaal
P. sturdeei		Bonin Is, S of Japan
P. tenuis (*papuanus*)		S Thailand – Java, Borneo, Philippines, Sulawesi, Timor, New Guinea, Bismarck Arch,; refs. 4.12, 72, 123, 128
P. wattsi		SE New Guinea; ref. 4.128
P. westralis		N, W Australia; (in *P. tenuis* ?); refs. 4.12, 72, 89, 123, 128

Subgenus *Vespadelus* (*Eptesicus*), (*Nycterikaupius*; ref. 4.124); refs. 4.123, 129

P. baverstocki		C Australia; ref. 4.128
P. caurinus		NW Australia
P. darlingtoni (*sagittula*)		SE Australia, Tasmania, Lord Howe I
P. douglasorum	Yellow-lipped bat	NW Australia
P. finlaysoni		W, C Australia; ref. 4.129
P. pumilus	Little bat	E Australia
P. regulus	King River bat	SW, S, SE Australia
P. troughtoni		E, NE Australia; ref. 4.129
P. vulturnus	Little forest bat	SE Australia, Tasmania

Subgenus *Perimyotis*; refs. 4.71, 123, 124

P. subflavus	Eastern pipistrelle	SE Canada – Guatemala, Honduras

Subgenus *Hypsugo*; ref. 4.123, 124

P. anchietae (*bicolor*)	Anchieta's pipistrelle	Angola, S Zaire, Zambia, Transvaal; ref. 4.123
P. anthonyi		N Burma
P. arabicus		N Oman; ref. 4.73
P. ariel	Desert pipistrelle	Egypt, Sudan, Israel; ref. 4.130
P. austenianus		Burma (in *P. savii* ?)
P. bodenheimeri	Bodenheimer's pipistrelle	Israel, SW Arabia, Oman, ? Socotra I
P. cadornae	Cadorna's pipistrelle	NE India – Thailand
P. curtatus		Enggano I, Sumatra; ref. 4.12
P. eisentrauti	Eisentraut's pipistrelle	Liberia, Ivory Coast, Cameroun, Kenya, Somalia; refs. 4.144, 145

? *P. helios*		Kenya, Sudan; ref. 4.123
P. hesperus	Western pipistrelle (Canyon bat)	Washington – C Mexico
P. imbricatus		Java, Borneo, Philippines, ? Sulawesi; refs. 4.12, 72
P. joffrei	Joffre's pipistrelle	Burma
P. kitcheneri		Borneo
P. lophurus		S Thailand, S Burma
P. macrotis		Malaya, Sumatra, Bali; ref. 4.14
P. musciculus	Least pipistrelle	Sierra Leone, Cameroun, C Zaire, Gabon
P. nanus	Banana bat	Sierra Leone – Somalia – S Africa; Madagascar
P. pulveratus	Chinese pipistrelle	S China, Hainan, Thailand
P. savii	Savi's pipistrelle	Iberia – Korea, Japan, Afghanistan; N Africa, Canary, Cape Verde Is
P. stenopterus		Malaya, Sumatra, Borneo, Philippines

Subgenus *Falsistrellus*; refs. 4.123, 128

P. affinis		India, NE Borneo, Yunnan
P. mackenziei		SW Australia; ref. 4.128
P. mordax		Java
P. petersi	Peters' pipistrelle	N Borneo – S Moluccas; (in *P. affinis* ?); ref. 4.131

Subgenus *Neoromicia* (*Eptesicus*); ref. 4.123

P. brunneus	Dark brown serotine	Liberia – C Zaire; ref. 4.144
P. capensis (*melckorum*)	Cape serotine	Africa S of Sahara; ref. 4.132
P. flavescens	Yellow serotine	Angola, Burundi; ref. 4.77
P. guineensis	Tiny serotine	Senegal – Ethiopia, NE Zaire, Kenya
P. rendalli	Rendall's serotine	Gambia – Somalia – Mozambique, Botswana
P. somalicus (*zuluensis*)	Somali serotine	Guinea-Bissau – Somalia – Kenya – S Africa; ? Madagascar
P. tenuipinnis	White-winged serotine	Guinea – Kenya – Angola

Subgenus *Arielulus*; ref. 4.123

P. circumdatus	Gilded black pipistrelle	N Burma, Malaya, Java
P. cuprosus		Borneo; ref. 4.14
P. societatis		Malaya

Scotozous

S. *dormeri*   India, Pakistan

Nyctalus

N. *aviator*   Korea, Japan, E China
N. *azoreum*   Azores Is; ref. 4.75
N. *lasiopterus*   Giant noctule   SW Europe – Caucasus,
                                                                            Iran; N Africa
N. *leisleri*   Lesser noctule   Madeira, Canary Is, N
                          (Leisler's bat)   Africa, W Europe –
                                                                            Kashmir
N. *montanus*   E Afghanistan – N India
N. *noctula*   Noctule   Britain, Algeria, W Europe
                                                                            – China, Japan, Taiwan,
                                                                            Vietnam, ? Malaya;
                                                                            ref. 4.76

Glischropus

G. *javanus*   Java; K
G. *tylopus*   Thick-thumbed pipistrelle   Burma – Sumatra, Borneo,
                                                                            Palawan, N Moluccas

Laephotis; Africa S of Sahara.

L. *angolensis*   Angola, S Zaire
L. *botswanae*   Botswana long-eared bat   NE Namibia – Zambia –
                                                                            Transvaal
L. *namibensis*   Namib long-eared bat   Namibia, SW Cape Prov.
L. *wintoni*   De Winton's long-eared bat   Ethiopia, Kenya, Tanzania

Philetor

P. *brachypterus*   New Guinea brown bat   E Nepal, Malaya – New
                                                                            Guinea, Bismarck Arch.,
                                                                            Philippines; refs. 4.80,
                                                                            81
Hesperoptenus

Subgenus *Hesperoptenus*

H. *doriae*   False serotine bat   Malaya, Borneo

Subgenus *Milithronycteris*

H. *blanfordi*   Blanford's bat   S Burma – Malaya, Borneo
H. *gaskelli*   Sulawesi; ref. 4.12
H. *tickelli*   Tickell's bat   India – Thailand,
                                                                            Andaman Is, Sri Lanka,
                                                                            S China
H. *tomesi*   Malaya, Borneo

Chalinolobus; Australia.

C. *dwyeri*   Large pied bat   C, S Queensland – E New
                                                                            South Wales

C. gouldii (*neocaledonicus*)	Gould's wattled bat	Australia, Tasmania, New Caledonia
C. morio	Chocolate wattled bat	S Australia, Tasmania
C. nigrogriseus	Hoary bat (Frosted bat)	SE New Guinea, Fergusson I, N Australia
C. picatus	Little pied bat	C, S Queensland – Victoria
C. tuberculatus	Long-tailed bat	New Zealand

Scotoecus; Africa – N India.

S. albigula		Somalia, Kenya, Zambia, Malawi, Angola; ref. 4.133, 145
S. albofuscus	Light-winged lesser house bat (Thomas's house bat)	Senegal – Tanzania, S Malawi, Mozambique, ? Zambia
S. hindei		Ghana – Somalia – Tanzania – Angola
S. hirundo		Senegal – Ethiopia; refs. 4.74, 84
S. pallidus		Pakistan, N India

Nycticeinops; ref. 4.123.

N. schlieffenii	Schlieffen's bat	Mauretania – Egypt – Namibia, Transvaal, Natal, SW Arabia

Scoteanax; (*Nycticeius*); ref. 4.85.

S. rueppellii	Rüppell's broad-nosed bat (Greater broad-nosed bat)	E Queensland, E New South Wales

Scotorepens; (*Nycticeius*); ref. 4.85.

S. balstoni (*influatus*)	Western broad-nosed bat	NW, N, C Australia
S. greyi (*caprenus*) (*aquilo*)	Little broad-nosed bat (Grey's bat)	Australia except SW, NE, SE
S. orion		SE Australia
S. sanborni	Little northern broad-nosed bat	N Australia, S, SE New Guinea

Little broad-nosed bat
(Scotorepens greyi)

Eptesicus; (*Vespertilio*); serotines, brown bats; Europe – E Asia, Africa, Alaska – Argentina.

Subgenus *Eptesicus*

E. bobrinskoi		SW USSR, NW Iran
E. bottae	Botta's serotine	NE Egypt – Arabia – Turkestan, Iran
E. brasiliensis (*andinus*)	Brazilian brown bat	S Mexico – Uruguay, SE Argentina; Trinidad, Tobago

E. chiriquinus	Chiriqui brown bat	Panama; (in *E. brasiliensis*?); ref. 4.56
E. demissus	Surat serotine	S Thailand
E. diminutus		E, SE Brazil – N Argentina, Uruguay
E. furinalis (*montosus*)	Argentine brown bat	Mexico – N Argentina; ref. 4.56
E. fuscus	Big brown bat	Alaska, S Canada – Colombia, Venezuela, Greater Antilles, Dominica; ref. 4.78
E. gobiensis (*centrasiaticus*) (*kashgaricus*)		S USSR, Gobi; ref. 4.135
E. guadeloupensis	Guadeloupe brown bat	Guadeloupe, Lesser Antilles
E. hottentotus	Long-tailed house bat	Namibia – Mozambique, S Africa
E. innoxius		W Ecuador, NW Peru
E. kobayashii		Korea; status uncertain
E. lynni	Lynn's brown bat	Jamaica
E. nasutus (*walli*)	Sind bat	SW Arabia – Pakistan; ref. 4.79
E. nilssonii	Northern bat	France, Norway – E. Siberia – Japan, N India
E. pachyotis	Thick-eared bat	Assam – N Thailand
E. platyops	Lagos serotine	Senegal, Nigeria; possibly subspecies of *E. serotinus*; ref. 4.134
E. serotinus	Serotine	England, NW Africa, Morocco, W Europe – Thailand, China, Korea
E. tatei		NE India; status uncertain

Subgenus *Rhinopterus*

E. floweri	Horn-skinned bat	Mali, S Sudan

Ia

I. io	Great evening bat	Assam – S China, Indochina, Thailand

Vespertilio

V. murinus	Particoloured bat	Scandinavia, Siberia – Iran, Afghanistan; vagrant
V. orientalis		E China, Honshu, Taiwan
V. superans		E Siberia, E China, Japan

Particoloured bat (Vespertilio murinus)

Histiotus; big-eared brown bats; Colombia – Chile, Argentina.

H. alienus		Brazil, Uruguay

H. macrotus		S Peru – NW Argentina, Chile
H. montanus		Venezuela, Colombia – Chile, Argentina, Uruguay
H. velatus		Brazil, Paraguay

Tylonycteris

T. pachypus	Bamboo bat (Lesser club-footed bat) (Flat-headed bat)	India, S China – Java, Lesser Sundas, Borneo, Philippines, Andaman Is
T. robustula	Greater club-footed bat (Flat-headed bat)	SW China – Java, Borneo, Philippines, Sulawesi, Timor; ref. 4.136

Mimetillus

M. moloneyi	Moloney's flat-headed bat	Sierra Leone – Ethiopia – Angola

Glauconycteris; Africa S of Sahara.

G. alboguttatus	Allen's striped bat	E Zaire, Cameroun
G. argentata	Silvered bat	Cameroun – Kenya – NE Angola, Tanzania
G. beatrix	Beatrix bat	Ivory Coast – Kenya
G. egeria	Bibundi bat	Cameroun, Uganda
G. gleni		Cameroun, Uganda
G. kenyacola		E Kenya; ref. 4.82
G. machadoi	Machado's butterfly bat	E Angola; possibly subspecies of *G. variegata*; ref. 4.83
G. poensis	Abo bat	Senegal – Uganda; ref. 4.74
G. superba	Pied bat	Ivory Coast, Ghana, NE Zaire
G. variegata	Butterfly bat	Senegal – Somalia – S Africa

Scotomanes

S. emarginatus		India
S. ornatus	Harlequin bat	NE India – S China, Vietnam

Scotophilus; Africa, SE Asia; classification in Africa uncertain; refs. 4.108, 138, 139, 144.

S. borbonicus		Reunion, Madagascar; refs. 4.17, 87, 138, 139
S. celebensis	Sulawesi yellow bat	N Sulawesi
S. dinganii	African yellow house bat	Senegal – Somalia – S Africa

S. heathii	Asiatic greater yellow house bat	Afghanistan – S China, Hainan, Vietnam, Sri Lanka
S. kuhlii	Asiatic lesser yellow house bat	Pakistan – Hainan – Timor, Aru Is, Philippines, Taiwan
S. leucogaster		Mauritania – Somalia, Aden; ? Namibia, Botswana; ref. 4.108
S. nigrita (*gigas*)	Greater brown bat (Giant yellow bat)	Senegal – S Sudan – Mozambique
S. nigritellus		Mali, Ivory Coast – Ethiopia; (in *S. borbonicus*?); ref. 4.108, 138, 139
S. nucella		Ghana, Uganda; ref. 4.109
S. nux		Sierra Leone – Kenya; ref. 4.138, 139
S. robustus		Madagascar; (in *S. dinganii*?)
S. viridis	Lesser yellow house bat	Tanzania, Angola, S Africa; (in *S. borbonicus*?); refs. 4.87, 108, 138, 139

Lasiurus; (*Nycteris*; ref. 4.40); hairy-tailed bats.

Subgenus *Lasiurus*

L. borealis (*pfeifferi*) (*degelidus*) (*minor*) (*brachyotis*)	Red bat	S Canada – C Chile, Argentina; Bahamas, Trinidad, Greater Antilles, Puerto Rico, Bermuda, Galapagos; ref. 4.40
L. castaneus	Tacarcuna bat	Panama
L. cinereus	Hoary bat	NC, S Canada – C Chile, C Argentina; Hawaii, Galapagos; Iceland, Orkney, Bermuda, (vagrant)
L. seminolus	Seminole bat	E USA, Bermuda, (vagrant)

Subgenus *Dasypterus*

L. ega	Southern yellow bat	SW USA – Argentina, Uruguay
L. egregius		Panama, Brazil
L. insularis		Cuba, I of Pines; refs. 4.39, 40, 88

L. intermedius	Northern yellow bat (Eastern yellow bat)	SE USA, N Mexico – Honduras

Barbastella; barbastelles.

B. barbastellus	Western barbastelle	England, France, Morocco, ? Senegal – Caucasus
B. leucomelas	Eastern barbastelle	Israel – Caucasus – N India, W China, Japan, ? NE Africa

Western barbastelle
(Barbastella barbastellus)

Plecotus; long-eared bats.

Subgenus *Plecotus*

P. auritus	Brown long-eared bat (Common long-eared bat)	Britain, France – NE China, Korea, Japan, ? N India
P. austriacus	Grey long-eared bat	Spain, S England – W China, NE India; Cape Verde Is, N Africa, Senegal
P. teneriffae		Canary Is; ref. 4.137

Subgenus *Corynorhinus*

P. mexicanus	Mexican big-eared bat	Mexico
P. rafinesquii	Rafinesque's big-eared bat (Eastern lump-nosed bat)	SE USA
P. townsendii	Townsend's big-eared bat (Lump-nosed bat)	SW Canada, W USA – Mexico; I

Idionycteris; (*Plecotus*).

I. phyllotis	Allen's big-eared bat	Nevada, Utah – W New Mexico – C Mexico

Nycticeius

N. humeralis (*cubanus*)	Evening bat (Twilight bat)	S Canada, C, SE USA – E Mexico, Cuba

Rhogeessa; C, S America.

R. genowaysi		Chiapas, Mexico; ref. 4.86
R. gracilis	Slender yellow bat	W Mexico
R. minutilla		N Venezuela, NE Colombia
R. mira	Least yellow bat	Michoacan, C Mexico
R. parvula	Little yellow bat	W Mexico
R. tumida	Central American yellow bat	E Mexico – Ecuador, S Brazil, Bolivia; Trinidad

Pallid bat
(Antrozous pallidus)

Baeodon; (*Rhogeessa*).

B. alleni	Allen's baeodon (Allen's yellow bat)	W, C Mexico

Otonycteris

O. hemprichii	Hemprich's long-eared bat	Algeria, Niger – Egypt – Afghanistan, Kashmir

Euderma

E. maculatum	Spotted bat (Pinto bat)	SW USA – N, C Mexico

Antrozous; (*Bauerus*).

A. pallidus (*koopmani*)	Pallid bat	Br. Colombia, W, C USA – C Mexico, Cuba; refs. 4.140, 141

Bauerus; (*Antrozous*); ref. 4.95.

B. dubiaquercus (*meyeri*)	Van Gelder's bat	Tres Marias Is, Veracruz; Mexico, Honduras

Subfamily Miniopterinae

Miniopterus; long-fingered bats, bent-winged bats; refs. 4.10, 12, 89, 90, 91.

M. australis (*solomonensis*)	Little long-fingered bat	Java, Borneo, Philippines – E Australia, New Caledonia, Loyalty Is; ref. 4.91, 92
M. fraterculus	Lesser long-fingered bat	Malawi – S Africa; Madagascar, ? Zambia
M. fuscus (*yayeyamae*)		Ryukyu Is
M. inflatus	Greater long-fingered bat	Liberia – Somalia, Madagascar, Zambia, Mozambique,
M. magnater (*macrodens*) (? *bismarckensis*)		S China, Malaya – Java, Borneo, New Guinea; ref. 4.91
M. medius	SE Asian long-fingered bat	S Thailand – Java, New Guinea, Bismarck Arch.
M. minor	Least long-fingered bat	Congo Rep. – Tanzania, Madagascar, Comoro Is
M. pusillus (*macrocneme*)		Nicobar Is, Thailand, Philippines, Sulawesi, Java – New Caledonia, Loyalty Is
M. robustior		Loyalty Is
M. schreibersii (*oceanensis*)	Schreiber's long-fingered bat	Africa, Madagascar, SW Europe – China, Japan, Philippines, Solomons, NW, N, E Australia; ref. 4.91

M. tristis
  (*propitristis*)
  (*celebensis*)
  (*insularis*)
  (*melanesiensis*)

Sulawesi, Philippines, New Guinea – Solomons, New Hebrides; refs. 4.90, 91

Subfamily Murininae

Murina; tube-nosed bats.

Subgenus *Murina*

M. aenea	Bronze tube-nosed bat	Malaya, Borneo; ref. 4.14
M. aurata	Little tube-nosed bat	Nepal – S China, Thailand; ref. 4.93
M. cyclotis	Round-eared tube-nosed bat	N India – S China – Malaya, Hainan – Borneo, Sri Lanka, (?) Philippines; ref. 4.12
M. florium	Flores tube-nosed bat	Lesser Sunda Is, Sulawesi, Moluccas; New Guinea; Bismarck Arch., N Australia; ref. 4.12, 143, 150
M. huttoni	Hutton's tube-nosed bat	Himalayas – S China, Malaya
M. leucogaster (*fusca*)	Greater tube-nosed bat	NE India, China – E Siberia, Japan
M. puta		Taiwan; status uncertain
M. rozendaali		N Borneo; ref. 4.14
M. silvatica		Japan; ref. 4.94
M. suilla (*balstoni*) (*canescens*)	Brown tube-nosed bat	Malaya – Java, Borneo
M. tenebrosa		Tsushima I, Japan
M. tubinaris		N Pakistan – Vietnam
M. ussuriensis		E, NE Siberia, Sakhalin, Kurile Is; Korea; ref. 4.142

Subgenus *Harpiola*

M. grisea	Peters' tube-nosed bat	NW India

Harpiocephalus

H. harpia	Hairy-winged bat	India – Vietnam, Taiwan, Sumatra, Java, Borneo, S Moluccas
H. mordax		Burma, ? Borneo; ref. 4.14

Hairy-winged bat
(Harpiocephalus harpia)

Subfamily Kerivoulinae

Kerivoula; woolly bats.

K. africana	Tanzanian woolly bat	Tanzania; extinct ?
K. agnella	Louisiade trumpet-eared bat	Sudest I, SE New Guinea; St Aignan's I (= Misima I), Louisiade Arch.
K. argentata	Damara woolly bat	S Kenya – Namibia – S Africa
K. cuprosa	Copper woolly bat	Ghana, S Cameroun, Kenya, Zaire
K. eriophora		Ethiopia; possibly conspecific with *K. africana*
K. flora		Borneo, ? Bali; Flores I, Lesser Sunda Is; ref. 4.143
K. hardwickii	Hardwick's forest bat	Sri Lanka, India – S China – Java, Bali, ? Sumba I, Lesser Sunda Is, Philippines, Sulawesi, Talaud Is; ref. 4.143
K. intermedia		Borneo, Malaya; ref. 4.14
K. lanosa	Lesser woolly bat	Liberia – Ethiopia – S Africa
K. minuta	Least forest bat	S Thailand, Malaya, Borneo; ref. 4.14
K. muscina	Fly River trumpet-eared bat	C New Guinea
K. myrella	Bismarck trumpet-eared bat	Wetar I, Lesser Sunda Is, Admiralty Is, Bismarck Arch.; ref. 4.143
K. papillosa	Papillose bat	NE India – Java, Borneo, Sulawesi; refs. 4.12, 143
K. pellucida	Clear-winged bat	Philippines, Malaya, Borneo, ? Sumatra, Java
K. phalaena	Spurrell's woolly bat	Liberia, Ghana, Cameroun, Congo Rep., Zaire
K. picta	Painted bat	Sri Lanka, S India – S China – Java, Lesser Sundas, Borneo, Ternate I, Ambon
K. smithii	Smith's woolly bat	Nigeria – E Zaire, Kenya
K. whiteheadi		S Thailand, Malaya, Borneo, Philippines

Phoniscus

P. aerosa	Dubious trumpet-eared bat	? SE Asia; origin and status uncertain
P. atrox	Groove-toothed bat	S Thailand, Malaya, Sumatra, Borneo; ref. 4.14
P. jagorii	Peters' trumpet-eared bat	Samar I, Philippines, Borneo, Java, Sulawesi
P. papuensis	Papuan trumpet-eared bat	SE New Guinea, E Australia

Subfamily Nyctophilinae

Nyctophilus; (*Lamingtona*); ref. 4.96.

N. arnhemensis	Arnhem Land long-eared bat	NW Australia
N. bifax	Northern long-eared bat	N Australia, New Guinea; ref. 4.152
N. geoffroyi	Lesser long-eared bat	Australia, Tasmania
N. gouldi	Gould's long-eared bat	W, E Australia; ref. 4.152
N. microdon	Small-toothed long-eared bat	SE New Guinea
N. microtis (lophorhina)	Papuan long-eared bat	SE New Guinea; ref. 4.96
N. timoriensis	Greater long-eared bat	NW, SW, SE Australia, New Guinea; Tasmania, ? Timor; ref. 4.97
N. walkeri	Pygmy long-eared bat	NW Australia

Pharotis

P. imogene	Big-eared bat	SE New Guinea

Subfamily Tomopeatinae

Tomopeas

T. ravus		NW Peru

Family Mystacinidae

New Zealand short-tailed bats; 2 species.

Mystacina; New Zealand; refs. 4.98, 99.

M. robusta	New Zealand greater short-tailed bat	Solomon I, Big South Cape I; probably extinct
M. tuberculata	New Zealand lesser short-tailed bat	New Zealand; Solomon I, Big South Cape I, Codfish I, Jacky Lee I; V

New Zealand lesser-short tailed bat (Mystacina tuberculata)

Family Molossidae

Free-tailed bats; *c.* 89 species; tropics and subtropics of Old and New Worlds; generic classification unstable; refs. 4.100, 101.

Tadarida; free-tailed bats, mastiff bats; refs. 4.100, 101.

Subgenus *Tadarida*; (*Rhizomops*).

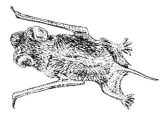

Brazilian free-tailed bat (Tadarida brasiliensis)

T. aegyptiaca	Egyptian free-tailed bat	Africa, Arabia, Iran – India, Sri Lanka
T. australis	Southern mastiff-bat (white-striped mastiff-bat)	C, S Australia
T. brasiliensis	Brazilian free-tailed bat	W, S USA – C Chile, Argentina, Bahamas, Antilles
T. fulminans	Madagascar large free-tailed bat	E Zaire, Kenya – Transvaal, Madagascar
T. kuboriensis	Small-eared mastiff bat	SE New Guinea; (in *T. australis*?); ref. 4.100
T. lobata	Kenya big-eared free-tailed bat	Kenya, Zimbabwe
T. teniotis	European free-tailed bat	S Europe, Madeira, Canary Is, N Africa – N India, China, Korea, Japan, Taiwan
T. ventralis (*africana*)	African giant free-tailed bat	Sudan, Ethiopia – Mozambique, Transvaal

Subgenus *Nyctinomops*

T. aurispinosa (*similis*)	Peale's free-tailed bat	E, W Mexico, Venezuela, Colombia, Peru, E Brazil
T. femorosacca	Pocketed free-tailed bat	SW USA – S Mexico
T. laticaudata (*gracilis*) (*yucatanica*) (*europs*) (*? espiritosantensis*)	Broad-tailed bat	NE Mexico – Venezuela – N Argentina, Cuba, Trinidad; refs. 4.60, 61
T. macrotis	Big free-tailed bat	SW British Colombia, E, C USA – Brazil, Uruguay, Paraguay, Greater Antilles

Subgenus *Mops*

T. condylura	Angola free-tailed bat	Senegal – Somalia – S Africa; Madagascar
T. congica	Medje free-tailed bat	Ghana, Nigeria, Cameroun, NE Zaire, Uganda

T. demonstrator	Mongalla free-tailed bat	Burkina Faso, Sudan, NE Zaire, Uganda
T. midas	Midas free-tailed bat	Senegal – Ethiopia – S Africa; Madagascar; SW Arabia
T. mops	Malayan free-tailed bat	Malaya, Sumatra, Borneo, ? Java
T. niangarae	Niangara free-tailed bat	NE Zaire
T. niveiventer	White-bellied free-tailed bat	Zaire – Angola, Zambia, Mozambique
T. sarasinorum (*lanei*)	Sulawesi mastiff-bat	Sulawesi, Mindanao I, Philippine Is; ref. 4.143
T. trevori		S Sudan, NE Zaire, Uganda

Subgenus *Xiphonycteris*; ref. 4.102

T. brachyptera		Uganda – Tanzania, Mozambique, Zanzibar
T. leonis	Sierra Leone free-tailed bat	Sierra Leone – E Zaire, Uganda, Bioco; (in *T. brachyptera*?); refs. 4.102, 144.
T. nanula	Dwarf free-tailed bat	Sierra Leone – Ethiopia, Kenya, Zaire
T. petersoni	Peterson's free-tailed bat	Ghana, Cameroun, ref. 4.102
T. spurrelli	Spurrell's free-tailed bat	Liberia – Rio Muni, Central African Rep. Zaire, Bioco; ref. 4.144
T. thersites	Railer bat	Liberia – S, E Zaire, Bioco, Mozambique, ? Zanzibar

Subgenus *Chaerephon*

T. aloysiisabaudiae	Duke of Abruzzi's free-tailed bat	Ghana, Gabon, N Zaire, Uganda, ? Ethiopia
T. ansorgei	Ansorge's free-tailed bat	Cameroun, Ethiopia – Angola, Transvaal
T. bemmelini	Gland-tailed free-tailed bat	Liberia – S Sudan – N Tanzania
T. bivittata	Spotted free-tailed bat	Ethiopia – Zambia – Mozambique
T. chapini	Chapin's free-tailed bat	W, NE Zaire, Uganda – Namibia, ? Ethiopia
T. gallagheri	Gallagher's free-tailed bat	C Zaire
T. jobensis	Northern mastiff-bat	New Guinea, N Australia, Solomons, Vanuatu, Fiji
T. johorensis	Dato Meldrum's bat	Malaya, Sumatra

T. major	Lappet-eared free-tailed bat	Ghana, Mali – S Sudan – Tanzania
T. nigeriae	Nigerian free-tailed bat	Ghana – Ethiopia, Zimbabwe, Namibia; SW Arabia
T. plicata	Wrinkle-lipped free-tailed bat	Sri Lanka, India – S China – Java, Bali, Borneo, Cocos-Keeling Is, Philippines, Hainan
T. pumila	Little free-tailed bat	Senegal – Somalia – Angola, S Africa, Madagascar, SW Arabia
T. pusillus		Aldabra I, Indian Ocean; (in *T. pumila?*)
T. russata	Russet free-tailed bat	Ghana, Cameroun, NE Zaire, Kenya

Mormopterus; (*Tadarida, Micronomus*); refs. 4.100, 101.

M. acetabulosus	Natal wrinkle-lipped bat (Natal free-tailed bat)	Ethiopia, Natal, Madagascar, Mauritius, Reunion
M. beccarii	Beccari's mastiff-bat	Moluccas, New Guinea, N Australia; ref. 4.143
M. doriae		Sumatra
M. jugularis	Peters' wrinkle-lipped bat	Madagascar
M. kalinowskii		Peru, N Chile
M. loriae	Little northern mastiff-bat	New Guinea, N Australia; (in *M. planiceps?*)
M. minutus	Little goblin bat	Cuba
M. norfolkensis	Eastern little mastiff-bat (Norfolk island bat)	SE Queensland, E New South Wales, ? Norfolk I
M. phrudus		Peru
M. planiceps	Litle mastiff bat	SW, C Australia

Sauromys; possibly a subgenus of *Mormopterus*; refs. 4.100, 101.

S. petrophilus	Roberts's flat-headed bat (Flat-headed free-tailed bat)	Namibia – Zimbabwe, Mozambique, S Africa

Platymops; possibly a subgenus of *Mormopterus*; refs. 4.100, 101.

P. setiger	Peters' flat-headed bat	S Sudan, S Ethiopia, Kenya

Otomops

O. formosus	Java mastiff-bat	Java; K
O. martiensseni	Martienssen's free-tailed bat (Giant mastiff-bat) (Large-eared free-tailed bat)	Ethiopia, Cent. African Rep. – Angola, Natal, Madagascar

O. papuensis	Big-eared mastiff-bat	SE New Guinea
O. secundus	Mantled mastiff-bat	NE New Guinea
O. wroughtoni	Wroughton's free-tailed bat	S India

Myopterus
M. albatus	Banded free-tailed bat	Ivory Coast, NE Zaire
M. daubentonii	Daubenton's free-tailed bat	Senegal; status uncertain
M. whitleyi	Bini free-tailed bat	Ghana – Zaire, Uganda

Cabreramops; (*Molossops*); ref. 4.103.
C. aequatorianus		W Ecuador

Molossops

Subgenus *Molossops*

M. neglectus		Surinam; ref. 4.104
M. temmincki		Venezuela, Colombia – N Argentina, Uruguay

Subgenus *Cynomops*

M. abrasus (*brachymeles*)		Guianas – Venezuela – Peru – N Argentina; ref. 4.105
M. greenhalli	Greenhall's dog-faced bat	W Mexico – Ecuador, NE Brazil, Trinidad; ref. 4.106
M. milleri		Peru; (in *M. planirostris*?); ref. 4.56
M. paranus		C Mexico, Venezuela, Brazil; (in *M. planirostris*?); refs. 4.53, 56, 104
M. planirostris	Southern dog-faced bat	Panama – Peru – N Argentina

Neoplatymops; possibly a subgenus of *Molossops*; ref. 4.100.
N. mattogrossensis		C Venezuela, S Guyana, NE, C Brazil

Eumops
E. auripendulus	Slouch-eared bat	S Mexico – N Argentina, Trinidad, Jamaica
E. bonariensis (*nanus*)	Peters' mastiff-bat	S Mexico – C Argentina, Uruguay
E. dabbenei		N Venezuela, N Colombia, N Argentina, Paraguay
E. glaucinus	Wagner's mastiff-bat	S Florida; C Mexico – N Argentina – SE Brazil; Cuba, Jamaica

E. hansae	Sanborn's mastiff-bat	Costa Rica – Peru, Brazil, Guyana
E. maurus	Guianian mastiff-bat	Guyana, Surinam
E. perotis (*trumbulli*) (? *gigas*)	Greater mastiff-bat	SW USA – N Mexico, Venezuela – N Argentina; ? Cuba
E. underwoodi	Underwood's mastiff-bat	S Arizona – Nicaragua; ref. 4.65
Promops		
P. centralis (*davisoni*) (*occultus*)	Thomas's mastiff-bat	W Mexico – Peru – N Argentina; Trinidad
P. nasutus (*pamana*)		N Argentina – Venezuela; Trinidad; ref. 4.100
Molossus		
M. ater	Black mastiff-bat	N Mexico – Guyana – Peru, N Argentina, Trinidad
M. barnesi		French Guiana, Brazil
M. bondae	Bonda mastiff-bat	Honduras – Ecuador, NW Venezuela
M. molossus (*coibensis*)	Pallas' mastiff-bat	N Mexico – N Argentina, Uruguay, Trinidad, Antilles; ref. 4.40
M. pretiosus (*macdougalli*)	Miller's mastiff-bat	S Mexico – Venezuela, Guyana; ref. 4.40
M. sinaloae	Allen's mastiff-bat	W Mexico – Costa Rica
M. trinitatis	Trinidad mastiff-bat	Costa Rica – Surinam, Trinidad; ref. 4.100
Cheiromeles		
C. parvidens		C Sulawesi, Philippines
C. torquatus	Hairless bat	Malaya – Java, Borneo, Philippines

ORDER PRIMATES

Primates; *c.* 201 species; S, C America, Africa, SE Asia; forest (savanna); mainly arboreal.

Family Cheirogaleidae

Mouse-lemurs, dwarf lemurs; *c.* 7 species; Madagascar; forest; ref. 5.4.

Microcebus; mouse-lemurs.

Lesser mouse-lemur (Microcebus murinus)

M. murinus	Lesser mouse-lemur (Grey mouse-lemur)	S, W Madagascar

| *M. rufus* | Russet mouse-lemur (Brown mouse-lemur) | E Madagascar |

Mirza; (*Microcebus*).
| *M. coquereli* | Coquerel's dwarf lemur | W Madagascar; R |

Cheirogaleus; dwarf lemurs.
| *C. major* | Greater dwarf lemur | N, E Madagascar |
| *C. medius* | Fat-tailed dwarf lemur | S, W Madagascar |

Allocebus
| *A. trichotis* | Hairy-eared dwarf lemur | NE Madagascar; E |

Phaner
| *P. furcifer* | Fork-marked lemur | N, W Madagascar; K |

Family Lemuridae

Large lemurs; *c.* 11 species; Madagascar; mainly forest; ref. 5.4.

Lemur
| *L. catta* | Ring-tailed lemur | S Madagascar |

Ringtailed lemur (Lemur catta)

Petterus; (*Lemur*); ref. 5.19.
P. coronatus	Crowned lemur	N Madagascar; K
P. fulvus (*albifrons*)	Brown lemur	Madagascar; K
P. macaco	Black lemur	NW Madagascar; V/E
P. mongoz	Mongoose-lemur	NW Madagascar, Comoro Is; V
P. rubriventer	Red-bellied lemur	E Madagascar; I

Hapalemur; gentle lemurs.
H. aureus	Golden lemur	SE Madagascar; ref. 5.10; I
H. griseus	Grey gentle lemur	E, N, WC Madagascar; K
H. simus	Broad-nosed gentle lemur	EC Madagascar; E

Varecia; (*Lemur*).
| *V. variegata* | Ruffed lemur | E Madagascar; I |

Lepilemur; sometimes treated as containing seven species; sometimes placed in separate family Lepilemuridae or Megalodapidae.
| *L. mustelinus* (*ruficaudatus*) (*septentrionalis*) (*dorsalis*) (*edwardsi*) (*leucopus*) (*microdon*) | Weasel-lemur, Sportive lemur | Madagascar; K |

*Verreaux's sifaka
(Propithecuverreauxi)*

*Aye-aye
(Daubentonia
madagascariensis)*

*Lesser bushbaby
(Galago senegalensis)*

Family Indriidae

Leaping lemurs; 5 species; Madagascar; forest; arboreal; ref. 5.4

Avahi; (*Lichanotus*).

A. laniger	Woolly lemur	E, NW Madagascar; K

Propithecus; sifakas.

P. diadema	Diadem sifaka	E, N Madagascar; V
P. tattersalli		NE Madagascar; ref. 5.20
P. verreauxi	Verreaux's sifaka	W, S Madagascar; K

Indri

I. indri	Indri	NE Madagascar; E

Family Daubentoniidae

One species.

Daubentonia

D. madagascariensis	Aye-aye	N Madagascar; forest; E

Family Lorisidae

Lorises, bushbabies; *c.* 15 species; SE Asia, Africa; forest, (savanna); arboreal, omnivorous.

Loris

L. tardigradus	Slender loris	S India, Sri Lanka

Nycticebus; slow lorises.

N. coucang	Slow loris	Assam – Java, Borneo, Philippines (Sulu Arch.)
N. pygmaeus	Pygmy slow loris	Indochina

Perodicticus

P. potto	Potto gibbon	Guinea – Zaire – Kenya; forest

Arctocebus

A. calabarensis	Angwantibo	R Niger – R Zaire; forest; K

Galago

G. alleni	Allen's bushbaby	R Niger – R Zaire; forest
G. granti	Grant's bushbaby	Mozambique, etc.
G. moholi	Southern lesser bushbaby	Angola – Mozambique – Tanzania
G. senegalensis	Northern lesser bushbaby	Senegal – Somalia – Tanzania; savanna

Otolemur; (*Galago*); greater bushbabies.

O. crassicaudatus		Angola – Kenya – Natal
O. garnettii		S Somalia – N Mozambique

Euoticus; (*Galago*); needle-clawed bushbabies; C Africa; forest.

E. elegantulus	Western needle-clawed bushbaby	R Niger – R Zaire
E. inustus	Eastern needle-clawed bushbaby	E Zaire

Galagoides; (*Galago*).

G. demidoff (*demidovii*)	Demidoff's galago	Senegal – Tanzania; forest
G. zanzibaricus	Zanzibar bushbaby	Kenya – Mozambique, Zanzibar

Family Tarsiidae

Tarsiers; 4 species; Sumatra, Borneo, Sulawesi; Philippines, etc.

Tarsius

T. bancanus	Western tarsier (Horsfield's tarsier)	Sumatra, Borneo
T. pumilus	Pygmy tarsier	C Sulawesi; ref. 5.11; I
T. spectrum	Spectral tarsier	Sulawesi
T. syrichta	Philippine tarsier	Mindanao, Philippines

Tarsier
(Tarsius bancanus)

Family Callithricidae

Marmosets, tamarins; *c.* 19 species; tropical S, C America; forest, (savanna);
arboreal; ref. 5.3.

Callithrix; (*Hapale*); marmosets.

C. argentata	Silvery marmoset
	    (Black-tailed marmoset)
		Brazil, Bolivia,
		    S of Amazon; (V)

C. humeralifer
    (*chrysoleuca*)
    (*santaremensis*)
	Santarem marmoset	R Madiera – R Tapajos,
		    Amazon Basin; V

C. jacchus
    (*flaviceps*)
    (*penicillata*)
	Common marmoset
	    (White-eared marmoset)
	    (White-headed
	    marmoset)
	    (Black-plumed
	    marmoset)
		E Brazil; (E)

Pygmy marmoset
(Cebuella pygmaea)

Cebuella

C. pygmaea	Pygmy marmoset	Upper Amazon Basin

Saguinus; (*Hapale, Leontocebus*); tamarins.

S. bicolor
    (*martinsi*)
	Bare-faced tamarin
	    (Martin's tamarin)
		NC Amazon Basin; I

S. fuscicollis	Saddle-back tamarin	Upper Amazon Basin
S. imperator	Emperor tamarin	W Brazil, E Peru, N Bolivia; I
S. inustus	Mottle-faced tamarin	NW Brazil, E Colombia
S. labiatus	White-lipped tamarin	C Amazon
S. leucopus	White-footed tamarin	N Colombia; V
S. midas (*tamarin*)	Red-handed tamarin (Negro tamarin)	N Brazil, Guianas
S. mystax	Moustached tamarin	N Peru, NW Brazil (S of Amazon)
S. nigricollis (*graellsi*)	Black and red tamarin	NW Brazil, etc.
S. oedipus (*geoffroyi*)	Cotton-top tamarin, Geoffroy's tamarin	Colombia, Panama; (E)
S. tripartitus (*fuscicollis*)		Peru; ref. 5.12

Leontopithecus; (*Leontocebus*, *Leontideus*); golden tamarins; ref. 5.5.

L. chrysomelas	Golden-headed tamarin	SE Bahia, Brazil; E
L. chrysopygus	Golden-rumped tamarin	Sao Paulo, Brazil; E
L. rosalia	Golden lion tamarin	SE Brazil; coastal forest; E

Callimico

| *C. goeldii* | Goeldi's marmoset | Upper Amazon Basin; R |

Common squirrel-monkey (Saimiri sciureus)

Family Cebidae

New-world monkeys; *c.* 45 species; S, C America; forest, arboreal; ref. 5.1.

Cebus; capuchin monkeys; Honduras – N Argentina.

C. albifrons	Brown pale-fronted capuchin	Upper Amazon, Venezuela, [Trinidad]
C. apella	Black-capped capuchin (Brown capuchin)	Colombia, Venezuela – N Argentina
C. capucinus	White-throated capuchin	Honduras – Colombia
C. olivaceus (*nigrivittatus*)	Weeper capuchin	Venezuela – mouth of Amazon

Aotus; night monkeys, douroucoulis; as many as 9 species have recently been recognized but their status as species or subspecies is uncertain; ref. 5.6.

| *A. azarae* (*infulatus*) (*miconax*) (*nancymai*) (*nigriceps*) | Southern night monkey | Amazon – Paraguay |
| *A. trivirgatus* (*brumbacki*) (*lemurinus*) (*vociferans*) | Northern night monkey | Panama – Amazon |

Callicebus; titis; S America; ref. 5.13..

C. brunneus		N Bolivia etc.
C. calligatus		Upper Amazon
C. cinerascens		S Amazon
C. cupreus		Upper Amazon
C. donacophilus		Bolivia, N Paraguay
C. dubius		C Amazon
C. hoffmannsi		C Amazon
C. modestus		NW Bolivia
C. moloch	Dusky titi	Lower Amazon (S of river)
C. oenanthe		Peru
C. olallae		NW Bolivia

Saimiri; squirrel monkeys; ref. 5.7.

S. boliviensis		Upper Amazon
S. oerstedii	Red-backed squirrel-monkey	Costa Rica, Panama
S. sciureus	Common squirrel-monkey	Colombia – Amazon Basin
S. ustus		C Brazil, S of Amazon
S. vanzolinii		C Amazon Basin; ref. 5.14

Pithecia; sakis; ref. 5.15.

P. aequatorialis	Equatorial saki	Ecuador, Peru
P. albicans	White saki	Central Amazon
P. irrorata	Bald-faced saki	S Amazon Basin
P. monachus	Monk saki	Upper Amazon Basin
P. pithecia	White-faced saki (Gold-faced saki)	Orinoco – Lower Amazon

Cacajao; uakaris; Amazon Basin.

C. calvus	White uakari (Bald uakari)	C Amazon; V
C. melanocephalus	Black-headed uakari	C Amazon – S Venezuela; V
C. rubicundus (calvus)	Red uakari	Upper Amazon Basin

Chiropotes; bearded sakis.

C. albinasus	White-nosed saki	S Amazon Basin; V
C. satanas	Black-bearded saki (Black saki)	Amazon – Guianas, Venezuela; (E)

Alouatta; howler monkeys; S Mexico – N Argentina.

A. belzebul	Black and red howler	Lower Amazon
A. caraya	Black howler	N Argentina – S Brazil
A. fusca	Brown howler	SE Brazil, Bolivia; I
A. palliata	Mantled howler	S Mexico – Ecuador
A. seniculus	Red howler	Colombia – mouth of Amazon – Bolivia
A. villosa (pigra)	Guatemalan howler	S Mexico – Guatemala; I

Ateles; spider monkeys; C Mexico – Bolivia.

A. belzebuth	Long-haired spider monkey	Colombia, NW Brazil, etc.; V
A. fusciceps	Brown-headed spider monkey	Panama – Ecuador; I
A. geoffroyi	Black-handed spider monkey	C Mexico – Colombia; V
A. paniscus	Black spider monkey	NE Brazil, Guianas – Bolivia; V

Brachyteles

B. arachnoides	Woolly spider monkey	SE Brazil; coastal forest; E

Lagothrix; woolly monkeys.

L. flavicauda	Yellow-tailed woolly monkey (Hendee's woolly monkey)	NE Peru; E
L. lagothricha	Common woolly monkey	C, Upper Amazon Basin; V

Savanna monkey (Cercopithecus aethiops)

Family Cercopithecidae

Old-world monkeys; *c.* 80 species; Africa, S, E Asia; forest, (savanna); ref. 5.1.

Subfamily Cercopithecinae

Macaca; (*Cynopithecus*); macaques; S, E Asia, (NW Africa); mainly forest; terrestrial, arboreal.

M. arctoides (*speciosa*)	Bear macaque (Stump-tailed macaque)	Burma, S China – Malaya
M. assamensis	Assam macaque	Nepal – Thailand
M. cyclopis	Taiwan macaque	Taiwan; V
M. fascicularis (*irus*)	Crab-eating macaque	S Burma – Java – Timor, Borneo, Philippines
M. fuscata (*speciosa*)	Japanese macaque	Japan (except Hokkaido)
M. maurus	Moor macaque	S Sulawesi; V
M. mulatta	Rhesus macaque	E Afghanistan – S, E China
M. nemestrina	Pigtail macaque	Assam – Sumatra, Borneo
M. nigra (*nigrescens*)	Celebes ape (Black ape)	NE Sulawesi
M. ochreata (*brunnescens*)	Booted macaque	SE Sulawesi; K
M. radiata	Bonnet macaque	S India
M. silenus	Liontail macaque	SW India; E
M. sinica	Toque macaque	Sri Lanka
M. sylvanus	Barbary ape	Morocco, N Algeria; V

M. thibetana	Père David's macaque (Tibetan stump-tailed macaque)	Sichuan, Fujian, China; K
M. tonkeana (*hecki*)	Tonkean macaque	C, N Sulawesi; (V)

Cercocebus; (*Lophocebus*); mangabeys; W, C, (E) Africa; forest.

C. albigena	White-cheeked mangabey	Cameroun – Kenya, Tanzania
C. aterrimus	Black mangabey (Crested mangabey)	Zaire, Angola; K
C. galeritus (*agilis*)	Agile mangabey (Tana River mangabey)	Zaire, etc., E Kenya, S Tanzania; V/E
C. torquatus (*atys*)	White-collared mangabey (Sooty mangabey)	Sierra Leone – Congo Rep.; V

Papio; savanna baboons; Africa except NW, (S Arabia); savanna.

P. anubis	Olive baboon	R Niger – Kenya
P. cynocephalus	Yellow baboon	Somalia – Mozambique – Angola
P. hamadryas	Hamadryas baboon	Ethiopia, Somalia, S Arabia; R
P. papio	Guinea baboon	Senegal, Guinea
P. ursinus	Chacma baboon	Zambia, Angola – S Africa

Mandrillus; (*Papio*); forest baboons; WC Africa; forest.

M. leucophaeus	Drill	SE Nigeria – Sanaga R (Cameroun); E
M. sphinx	Mandrill	Sanaga R – Gabon; V

Theropithecus

T. gelada	Gelada	Ethiopia; montane; R

Cercopithecus; guenons; Africa S of Sahara; forest, (savanna).

C. aethiops (*pygerythrus*) (*tantalus*) (*sabaeus*)	Savanna monkey (Green monkey) (Vervet) (Grivet)	Senegal – Somalia – S Africa; savanna
C. ascanius	Schmidt's guenon, etc.	W Kenya – Zambia, Angola
C. campbelli	Campbell's monkey	Gambia – Ghana
C. cephus	Moustached monkey	Cameroun – Gabon, etc.
C. denti (*wolfi*)	Dent's monkey	NE, E Zaire
C. diana	Diana monkey	Sierra Leone – Ghana; V
C. dryas	Dryas monkey	C Zaire
C. erythrogaster	Red-bellied monkey	SW Nigeria; E
C. erythrotis	Red-eared monkey	Nigeria, Cameroun; E
C. hamlyni	Owl-faced monkey	E Zaire, etc.; V
C. lhoesti	L'Hoest's monkey	SW Uganda, NE Zaire, etc.; V

C. mitis
  (*albogularis*)
	Diademed monkey
	  (Sykes' monkey)
		Somalia – Zaire – S Africa

C. mona	Mona monkey	Ghana – Cameroun
C. neglectus	De Brazza's monkey	Cameroun – Ethiopia – Angola
C. nictitans	Greater white-nosed monkey	Guinea – Zaire
C. petaurista	Lesser white-nosed monkey	Guinea – Togo
C. pogonias	Crowned guenon	SE Nigeria – N, W Zaire, etc.
C. preussi (*lhoesti*)	Preuss's monkey	W Cameroun, etc.; E
C. salongo	Zaire Diana monkey	C Zaire; ref. 5.8; K
C. solatus	Sun-tailed monkey	Gabon; ref. 5.16; V
C. wolfi	Wolf's monkey	C Zaire

Miopithecus; (*Cercopithecus*).

M. talapoin	Talapoin	Gabon – W Angola; forest

Allenopithecus; (*Cercopithecus*).

A. nigroviridis	Allen's swamp monkey
	  (Blackish-green guenon)
		Zaire, Congo Rep.; forest;
		  K

Erythrocebus; (*Cercopithecus*).

E. patas	Patas monkey
	  (Red monkey)
		Senegal – Ethiopia –
		  Tanzania; savanna,
		  steppe

Subfamily Colobinae; see ref. 5.9 for alternative classification.

Colobus; (*Procolobus*); colobus monkeys, W, C, E Africa; forest.

Guereza
(Colobus guereza)

C. angolensis	Angolan colobus	Angola – Kenya

C. badius
  (*rufomitratus*)
	Red colobus
	  (Bay colobus)
		Gambia – Tanzania; E/V

C. guereza	Guereza
	  (Eastern black-and-white
	  colobus)
		E Nigeria – Ethiopia –
		  Tanzania

C. kirkii	Kirk's colobus	Zanzibar

C. polykomos	King colobus)
	  (Western black-and-
	  white colobus)
		Gambia – Togo

C. satanas	Black colobus	Cameroun – Gabon; E

Procolobus; (*Colobus*).

P. verus	Olive colobus	Sierra Leone – Nigeria; forest; V

Pygathrix; (*Rhinopithecus*).

P. avunculus	Tonkin snub-nosed monkey	N Vietnam; E

P. brelichi	Brelich's snub-nosed monkey	Guizhou, S China; close to *P. roxellana*
P. nemaeus (*nigripes*)	Douc langur	Indochina, Hainan; E
P. roxellana (*bieti*)	Chinese snub-nosed monkey (Golden monkey) (Snow monkey)	Yunnan, Sichuan, S China; E/V

Simias; (*Nasalis*).

S. concolor	Pig-tailed langur	Mentawai Is (Sumatra); E

Nasalis

N. larvatus	Proboscis monkey	Borneo; V

Presbytis; (*Semnopithecus*; *Trachypithecus*); leaf monkeys, surelis; SE Asia; forest.

P. aurata		Java – Lombok; ref. 5.17
P. comata (*hosei, aygula*) (*thomasi*)	Sunda leaf monkey (Grizzled leaf monkey) (Ebony leaf monkey)	Sumatra, Java, Borneo; E
P. cristata	Silvered leaf monkey	Indochina – Sumatra
P. entellus	Hanuman langur	India, etc., Sri Lanka
P. francoisi (*delacouri*)	François' monkey	S China, Indochina; E
P. frontata	White-fronted leaf monkey	Borneo
P. geei	Golden leaf monkey	Assam, Bhutan; R
P. johnii	Nilgiri langur (John's leaf monkey)	SW India; (in *P. vetulus?*); E
P. melalophos (*femoralis*)	Banded leaf monkey (Mitred leaf monkey)	Thailand – Sumatra, Borneo
P. obscura	Dusky leaf monkey (Spectacled leaf monkey)	S Thailand, Malaya
P. phayrei	Phayre's leaf monkey	Burma, Thailand, etc.
P. pileata	Capped leaf monkey	Assam, Burma
P. potenziani	Mentawai leaf monkey	Mentawai Is (Sumatra); E
P. rubicunda	Maroon leaf monkey	Borneo
P. vetulus (*senex*)	Purple-faced leaf monkey	Sri Lanka

Family Hylobatidae

Gibbons; 9 species; SE Asia; forest; ref. 5.18.

Hylobates; (*Nomascus, Symphalangus*); gibbons; SE Asia.

H. agilis	Agile gibbon	Malaya, Sumatra, Borneo
H. concolor	Crested gibbon (Black gibbon)	Indochina, Hainan; V
H. hoolock	Hoolock gibbon	Assam, Burma, Yunnan; V
H. klossii	Kloss's gibbon (Dwarf siamang)	Mentawai Is (Sumatra); E

Common gibbon
(*Hylobates lar*)

H. lar	Common gibbon	S Burma – N Sumatra
H. moloch	Javan gibbon	Java; V
H. muelleri	Müller's gibbon	Borneo
H. pileatus	Pileated gibbon	SE Thailand, etc.; V
H. syndactylus	Siamang	Malaya, Sumatra

Family Pongidae

Apes; 4 species; W, C Africa, SE Asia; forest.

Pongo

P. pygmaeus	Orang-utan	Sumatra, Borneo; E

Pan; chimpanzees.

P. paniscus	Pygmy chimpanzee (Bonobo)	Zaire (S of Zaire R.); V
P. troglodytes	Chimpanzee	Guinea – Zaire – Uganda, Tanzania; E/V

Gorilla

G. gorilla	Gorilla	SE Nigeria – W Zaire; E Zaire, etc.; E/V

Chimpanzee
(Pan troglodytes)

Family Hominidae

Homo

H. sapiens	Man	Worldwide

ORDER CARNIVORA

Carnivores; *c.* 235 species; worldwide; terrestrial, arboreal, aquatic carnivores and omnivores.

Family Canidae

Dogs, foxes; *c.* 35 species; Eurasia, Americas, Africa, [Australia].

Canis; dogs, jackals; Africa, Eurasia, N, C America; desert – open forest.

C. adustus	Side-striped jackal	Senegal – Somalia – S Africa; savanna, steppe
C. aureus	Golden jackal	Senegal – Thailand, Sri Lanka; steppe – open forest
C. latrans	Coyote	Alaska, Canada, USA (except SE) – Costa Rica; desert – grassland
C. lupus	Wolf	Palaearctic (except N Africa), India, Alaska, Canada, Mexico, † USA; tundra, steppe, open forest; ancestor of domestic dogs, *C. familiaris*; V

Side-striped jackal
(Canis adustus)

C. mesomelas	Black-backed jackal	S, E Africa – Cameroun; savanna, steppe
C. rufus (*niger*)	Red wolf	SE Texas, etc.; E
C. simensis	Simian jackal (Simian fox)	Ethiopia; high montane grassland; E

Alopex; (*Vulpes*).

A. lagopus	Arctic fox	Eurasia, N America; tundra

Vulpes; (*Fennecus*, *Urocyon*); foxes; Africa, Eurasia, N America; desert – forest.

V. bengalensis	Bengal fox	India, Pakistan, Nepal; steppe – open forest; K
V. cana	Blanford's fox	Afghanistan etc.; montane steppe
V. chama	Cape fox	S Africa; steppe
V. corsac	Corsac fox	S Russia – Manchuria; steppe
V. ferrilata	Tibetan fox	Tibet, etc.; steppe
V. macrotis	Kit fox	SW USA, N Mexico; desert, steppe; (in *V. velox*?)
V. pallida	Pale fox	Senegal – Somalia; desert
V. rueppellii	Sand fox	Morocco – Afghanistan; desert, steppe
V. velox	Swift fox	C USA; grassland
V. vulpes (*fulva*)	Red fox	Palaearctic – Indochina, Canada, USA, [Australia]; forest; includes domesticated forms, e.g. silver fox
V. zerda	Fennec fox	Morocco – Arabia; desert

Urocyon; (*Vulpes*); grey foxes; N America.

U. cinereoargenteus	Grey fox	S Canada – Venezuela
U. littoralis	Island grey fox	Santa Barbara Is, California

Dusicyon; (*Atelocyon*, *Cerdocyon*, *Lycalopex*); S America; forest – desert; classification provisional.

D. australis†	Falkland Island wolf	Falkland Is (extinct)
D. culpaeus (*culpaeolus*)	Colpeo fox	Ecuador – Tierra del Fuego
D. griseus (*fulvipes*)	Argentine grey fox	Argentina, Chile
D. gymnocercus	Pampas fox	Argentina – Paraguay
D. microtis	Small-eared zorro	Amazon, Orinoco basins; forest; K

D. sechurae	Sechura fox	NW Peru, SW Ecuador; desert
D. thous	Common zorro (Crab-eating fox)	Colombia – N Argentina; savanna, open forest
D. vetulus	Hoary fox	C, S Brazil

Nyctereutes
N. procyonoides	Raccoon-dog	Indochina – SE Sibera, Japan, [E, C Europe]; forest

Chrysocyon
C. brachyurus	Maned wolf	E Brazil – N Argentina; grassland; V

Speothos
S. venaticus	Bush dog	Panama – SE Brazil, E of Andes; forest; V

Cuon
C. alpinus	Dhole (Red dog)	Altai, Pamirs – E Siberia – Java; forest; V

Lycaon
L. pictus	Hunting dog	Ivory Coast – Somalia – S Africa; steppe, savanna; V

Otocyon
O. megalotis	Bat-eared fox	S Africa – S Zambia; Tanzania – Ethiopia; grassland, steppe

Family Ursidae

Bears; 8 species; Eurasia, N, (S) America; mainly forest; omnivores, (carnivores).

Tremarctos
T. ornatus	Spectacled bear	Venezuela – Bolivia; forest; V

Selenarctos; (*Ursus*).
S. thibetanus	Asiatic black bear	Afghanistan – Indochina – E Siberia, Japan; forest; (E)

Ursus
U. americanus	American black bear	C Mexico – Alaska; forest
U. arctos (*horribilis*)	Brown bear (Grizzly bear)	Eurasia N of Himalayas, N America; mainly forest

Sun bear
(Helarctos malayanus)

Thalarctos; (*Ursus*).

T. *maritimus*   Polar bear   Arctic Ocean; V

Helarctos

H. *malayanus*   Sun bear   Burma – Sumatra, Borneo;
                                                   forest

Melursus

M. *ursinus*   Sloth bear   India, Sri Lanka; forest; I

Ailuropoda

A. *melanoleuca*   Giant panda   Sichuan etc., S China;
                                                   bamboo forest; ref. 6.4;
                                                   R

Family Procyonidae

Raccoons etc.; *c.* 13 species; S, C, (N) America; forest; omnivores.

Bassariscus; (*Jentinkia*).

B. *astutus*   Ring-tailed cat   Oregon – S Mexico
                        (Cacomistle)
B. *sumichrasti*   Central American   S Mexico – Panama
                        cacomistle

Common raccoon
(*Procyon lotor*)

Procyon; raccoons; insular forms in the W Indies and Mexico are sometimes treated as distinct species but are likely to represent *P. lotor*.

P. *cancrivorus*   Crab-eating raccoon   Costa Rica – N Argentina
P. *lotor*   Common raccoon   S Canada – Panama,
  (*gloveralleni*)   [Bahamas, Guadeloupe];
  (*insularis*)   (K)
  (*minor*)
  (*pygmaeus*)

Nasua; coatis, coatimundis; S, C, (N) America.

N. *nasua*   Coati   S USA – S America except
  (*narica*)   Patagonia
N. *nelsoni*   Cozumel coati   Cozumel I, Yucatan,
                                                   Mexico; K

Nasuella

N. *olivacea*   Little coatimundi   Venezuela – Ecuador

Potos

P. *flavus*   Kinkajou   E Mexico – S Brazil; forest

Bassaricyon; olingos; C, northern S America; possibly all one species.
B. *alleni*   Ecuador, Peru
B. *beddardi*   Guyana, etc.
B. *gabbii*   Bushy-tailed olingo   Nicaragua – Ecuador,
                                                   Venezuela

B. lasius	Harris' olingo	Costa Rica
B. pauli	Chiriqui olingo	W Panama

Family Ailuridae

One species.

*Red panda
(Ailurus fulgens)*

Ailurus

A. fulgens	Lesser panda (Red panda)	Nepal – W Burma – Sichuan; K

Family Mustelidae

Weasels, etc., *c.* 64 species; Americas, Eurasia, Africa; forest – desert, (aquatic); carnivores, (omnivores).

*American mink
(Mustela vison)*

Mustela; (*Grammogale, Lutreola, Putorius*); weaseis; Americas, Eurasia, (N Africa); forest, steppe, tundra.

M. africana	Amazon weasel	Amazon Basin (sic)
M. altaica	Mountain weasel	Himalayas – Altai – Korea
M. amurensis	Amur polecat	N Manchuria; ref. 6.3
M. erminea	Stoat (Ermine)	Europe – E Siberia, Japan, Alaska – N Greenland – N USA; [New Zealand]; forest, tundra
M. eversmanni	Steppe polecat	E Europe – Manchuria – Tibet; grassland, steppe
M. felipei		Colombia; montane
M. frenata	Long-tailed weasel	S Canada – Venezuela – Peru
M. kathiah	Yellow-bellied weasel	W Himalayas – S China
M. lutreola	European mink	W Siberia, E Europe, (W Europe); riversides, marshes; V
M. lutreolina		Java; K
M. nigripes	Black-footed ferret	Alberta – N Texas; grassland; E
M. nivalis (*rixosa*) (*minuta*)	Weasel (Lesser weasel)	N Africa, W Europe – E Siberia; Japan, Alaska – NE USA, [New Zealand]
M. nudipes	Malaysian weasel	Malaya, Sumatra, Borneo
M. putorius	Polecat	Europe; forest; probable ancestor of domestic ferret, *M. furo*
M. sibirica	Siberian weasel (Kolinsky)	Siberia – Himalayas, Thailand; Japan
M. strigidorsa	Back-striped weasel	Nepal – Thailand

| *M. vison* | American mink | Canada, USA, [Iceland, N, C Europe, Siberia]; rivers, lakes; includes domesticated mink |

Vormela

| *V. peregusna* | Marbled polecat | SE Europe – W China; steppe, grassland; (V) |

Martes; (*Charronia*); martens; Eurasia, N America; forest; arboreal carnivores.

M. americana	American marten	Canada, N USA
M. flavigula (*gwatkinsi*)	Yellow-throated marten	E Siberia – Java, Borneo; S India; (I)
M. foina	Beech marten	W Europe – Himalayas, Altai
M. martes	Pine marten	W Europe – W Siberia
M. melampus	Japanese marten	Japan (except Hokkaido), S Korea; (I)
M. pennanti	Fisher	Canada, N USA
M. zibellina	Sable	Siberia, Hokkaido

Eira; (*Galera, Tayra*).

| *E. barbara* | Tayra | NE Mexico – Argentina |

Galictis; (*Grison, Grisonella*); grisons; S Mexico – Brazil.

| *G. cuja* | Little grison | Peru – Uruguay – Chile |
| *G. vittata* | Greater grison | S Mexico – Peru – E Brazil |

Lyncodon

| *L. patagonicus* | Patagonian weasel | Argentina, Chile; grassland |

Ictonyx; (*Zorilla*).

| *I. striatus* | Zorilla (Striped polecat) | Senegal – Ethiopia – S Africa; steppe, savanna |

Poecilictis

| *P. libyca* | Saharan striped weasel | Sahara, etc.; desert, steppe |

Poecilogale

| *P. albinucha* | White-naped weasel (Striped weasel) | S Africa – Zaire, Uganda; savanna |

Gulo

| *G. gulo* (*luscus*) | Wolverine (Glutton) | Scandinavia, Siberia, Alaska, Canada, W USA; con. forest, tundra; V |

Mellivora

| *M. capensis* | Ratel (Honey badger) | N India – Arabia, Africa S of Sahara; steppe, savanna |

Eurasian badger
(Meles meles)

Meles

M. meles	Eurasian badger	Europe – Japan – S China; forest, grassland

Arctonyx

A. collaris	Hog-badger	N China – NE India – Sumatra

Mydaus; (*Suillotaxus*); stink badgers.

M. javanensis	Sunda stink badger	Sumatra, Java, Borneo
M. marchei	Palawan stink badger	Palawan, Calamian Is, Philippines

Taxidea

T. taxus	American badger	SW Canada – C Mexico; grassland, steppe

Melogale; (*Helictis*); ferret-badgers; SE Asia, grassland, open forest.

M. everetti	Everett's ferret-badger	Borneo; K
M. moschata	Chinese ferret-badger	NE India – S China – Indochina; Java, Borneo
M. personata (*orientalis*)	Burmese ferret-badger	Nepal – Indochina; Java; (K)

Mephitis

M. macroura	Hooded skunk	SW USA – Costa Rica
M. mephitis	Striped skunk	S Canada – N Mexico

Spilogale; spotted skunks; N, C America.

S. putorius (*angustifrons*) (*gracilis*)	Spotted skunk	SE, C USA – Costa Rica
S. pygmaea	Pygmy spotted skunk	W, SW Mexico

Conepatus; hog-nosed skunks; C, S America.

C. chinga (*rex*)		Chile – Peru, S Brazil
C. humboldti (*castaneus*)		Patagonia – Paraguay
C. leuconotus	Eastern hog-nosed skunk	E Texas, E Mexico
C. mesoleucus	Hog-nosed skunk	S USA – Nicaragua; (I)
C. semistriatus	Striped hog-nosed skunk	S Mexico – Peru – E Brazil

Lutra; (*Lontra, Lutrogale*); river otters; Eurasia, Americas, Africa; rivers, lakes, (sea-coast); *.

L. canadensis	Canadian otter	Canada, USA
L. felina		N Peru – S Chile; coastal; V
L. longicaudis (*annectens*)		C Mexico – Uruguay; (V)

L. lutra	European otter	Eurasia, N Africa, Sri Lanka, Taiwan, Sumatra, Java; (V)
L. maculicollis	Spotted-necked otter	Liberia – Ethiopia – S Africa
L. perspicillata	Smooth-coated otter	Iraq; India – Sumatra, Borneo; K
L. provocax		Chile, S Argentina; V
L. sumatrana	Hairy-nosed otter	Indochina – Java, Borneo; K

Pteronura

P. brasiliensis	Giant otter	Venezuela – Argentina; rivers; V

Aonyx; (*Amblonyx, Paraonyx*); clawless otters; Africa, SE Asia.

A. capensis	African clawless otter	Senegal – Ethiopia – S Africa
A. cinerea	Oriental small-clawed otter	India, S China – Java, Borneo, Palawan; K
A. congica (*microdon*)	Zaire clawless otter	Congo basin, etc.; forest

Enhydra

E. lutris	Sea otter	Bering Sea – California; rocky coasts

Family Viverridae

Civets, etc.; *c.* 35 species; S Asia, Africa, Madagascar; forest – steppe; mainly omnivores.

African civet
(Viverra civetta)

Poiana

P. richardsonii	African linsang	Sierra Leone – N Zaire; forest; (I)

Genetta; genets; Africa, (Arabia, SW Europe); forest, savanna; arboreal; classification very provisional; ref. 6.1.

G. abyssinica	Abyssinian genet	Ethiopia – Somalia; K
G. angolensis (*mossambicus*)	Angolan genet	Angola – Mozambique
G. cristata (*servalina*)		Cameroun – SE Nigeria
G. genetta (*felina*)	Small-spotted genet	Africa, SW Europe, Arabia – Asia Minor; savanna; (R)
G. johnstoni	Johnston's genet	Liberia; K
G. pardina	Pardine genet	Gambia – Cameroun; forest
G. rubiginosa (*maculata*)	Rusty-spotted genet	Senegal – Somalia – Transvaal; savanna

G. servalina	Servaline genet	Cameroun – Kenya; forest
G. thierryi	Hausa genet	Senegal – Chad; guinea savanna
G. tigrina	Large-spotted genet	S Africa
G. victoriae	Giant genet	Uganda, NE Zaire

Viverricula

V. indica (*malaccensis*)	Small Indian civet	India – S China – Java; Sri Lanka; [Madagascar, Zanzibar]

Osbornictis

O. piscivora	Congo water civet (Aquatic genet)	NE Zaire; lowland forest

Viverra; (*Civettictis*); civets; Africa, SE Asia.

V. civetta	African civet	Senegal – Somalia – Transvaal
V. megaspila	Large-spotted civet	India – Indochina – Malaya; (E)
V. tangalunga	Malay civet	Malaya, Sumatra, Borneo, Philippines
V. zibetha	Large indian civet	India – S China – Malaya

Prionodon; (*Pardictis*); linsangs; SE Asia.

P. linsang	Banded linsang	Thailand – Java, Borneo
P. pardicolor	Spotted linsang	Nepal – Indochina

Nandinia

N. binotata	African palm civet (Tree civet)	Guinea – S Sudan – Mozambique; forest

Arctogalidea

A. trivirgata	Three-striped palm civet	Assam – Java, Borneo; (I)

Paradoxurus

P. hermaphroditus (*philippinensis*)	Common palm civet (Toddy cat)	India – S China – Java, Borneo, Timor, Sulawesi, Philippines, Sri Lanka
P. jerdoni	Jerdon's palm civet	S India; I
P. zeylonensis	Golden palm civet	Sri Lanka

Paguma

P. larvata	Masked palm civet	Himalayas – China – Sumatra, Borneo, Taiwan

Macrogalidea

M. musschenbroekii	Brown palm civet	NE Sulawesi; R

Arctictis
A. binturong (whitei)	Binturong	Burma – Java, Borneo, Palawan

Fossa
F. fossa (fossana)	Malagasy civet	NW Madagascar; forest; V

Hemigalus; (Diplogale).
H. derbyanus	Banded palm civet	Malaya, Sumatra, Borneo
H. hosei	Hose's civet	Borneo

Chrotogale
C. owstoni	Owston's palm civet	Indochina; K

Cynogale
C. bennettii	Otter-civet	Indochina – Sumatra, Borneo; K

Eupleres
E. goudotii (major)	Falanouc	N Madagascar; forest; V

Cryptoprocta; family allocation dubious.
C. ferox	Fossa	Madagascar; forest, savanna; K

Family Herpestidae

Mongooses; c. 39 species; Africa, S Asia; forest – steppe; carnivores and omnivores; sometimes included in the Viverridae.

Galidea
G. elegans	Ring-tailed mongoose	Madagascar; forest

Galidictis
G. fasciata (striata)	Broad-striped mongoose	E Madagascar; forest; I
G. grandidiensis		SW Madagascar; ref. 6.5; K

Banded mongoose
(Mungos mungo)

Mungotictis
M. decemlineata (lineata) (substriatus)	Narrow-striped mongoose	W, SW Madagascar; savanna; V

Salanoia
S. concolor (olivacea)	Salano	NE Madagascar; forest; K

Suricata

S. suricatta	Meerkat (Suricate)	S Africa – S Angola

Herpestes; (*Xenogale*); common mongooses; Africa, S Asia, (SW Europe); forest – steppe; terrestrial.

H. auropunctatus (*javanicus*) (*palustris*)	Small Indian mongoose	E Arabia – Malaya; [Hawaii, W Indies, Guyana, etc.]
H. brachyurus	Short-tailed mongoose	Malaya, Sumatra, Borneo, Palawan
H. edwardsii	Indian grey mongoose	Arabia – Assam, Sri Lanka; [Malaya, Ryukyu Is, Mauritius]
H. fuscus	Indian brown mongoose	S India, Sri Lanka; (in *H. brachyurus*?)
H. hosei	Hose's mongoose	Borneo; (in *H. brachyurus*?)
H. ichneumon	Egyptian mongoose (Large grey mongoose)	Iberia; Arabia, etc.; Africa S of Sahara; savanna
H. javanicus	Javan mongoose	Indochina, Malaya, Java
H. naso	Long-nosed mongoose	Zaire – SE Nigeria; lowland forest
H. semitorquatus	Collared mongoose	Sumatra, Borneo
H. smithii	Ruddy mongoose	India, Sri Lanka
H. urva	Crab-eating mongoose	Nepal – S China – Thailand
H. vitticollis	Stripe-necked mongoose	India, Sri Lanka

Helogale; dwarf mongooses; E, S Africa; savanna.

H. hirtula		Ethiopia, Somalia, Kenya
H. parvula		Ethiopia – Natal – Namibia

Dologale

D. dybowskii	Pousargues' mongoose	Cent. Afr. Rep. – W Uganda; savanna

Galerella; (*Herpestes*); Africa; ref. 6.6.

G. nigrita		C, N Namibia
G. pulverulenta	Cape grey mongoose	S Africa – S Angola; steppe
G. sanguinea	Slender mongoose	Africa S of Sahara; savanna, steppe

Atilax

A. paludinosus	Marsh mongoose (Water mongoose)	Africa S of Sahara; rivers, marshes

Mungos

M. gambianus	Gambian mongoose	Gambia – Nigeria; savanna

M. mungo	Banded mongoose	Africa S of Sahara

Crossarchus; W, C Africa; forest; ref. 6.2.

C. alexandri		N Zaire, Uganda
C. ansorgei		C Zaire, N Angola
C. obscurus	Cusimanse (Long-nosed mongoose)	Sierra Leone – Ghana
C. platycephalus		Benin – Cameroun

Liberiictis

L. kuhni	Liberian mongoose	Liberia; E

Ichneumia

I. albicauda	White-tailed mongoose	Africa S of Sahara, S Arabia; savanna, open forest

Bdeogale; (*Galeriscus*).

B. crassicauda	Bushy-tailed mongoose	Kenya – Mozambique; (E/I)
B. jacksoni	Jackson's mongoose	Kenya, Uganda; K
B. nigripes	Black-legged mongoose	SE Nigeria – Zaire – N Angola

Rhynchogale

R. melleri	Meller's mongoose	S Zaire – Transvaal

Cynictis

C. penicillata	Yellow mongoose	S Africa – S Angola

Paracynictis

P. selousi	Grey meerkat (Selous' mongoose)	Mozambique – Angola

Family Hyaenidae

Hyaenas; 4 species; Africa, SW Asia; savanna – desert; terrestrial carnivores and scavengers.

Proteles; sometimes placed in separate family Protelidae.

P. cristatus	Aardwolf	S, E Africa; steppe

Crocuta

C. crocuta	Spotted hyaena	Africa S of Sahara; steppe, savanna

Hyaena

H. brunnea	Brown hyaena	S Africa – Zimbabwe; V
H. hyaena	Striped hyaena	Senegal – Tanzania – Turkestan – India; steppe, desert; (E)

Spotted hyaena
(Crocuta crocuta)

Family Felidae

Cats; *c.* 36 species; Americas, Eurasia, Africa; forest – desert; carnivores; generic classification very unstable.

Serval
(Felis serval)

Felis; (*Badiofelis, Herpailurus, Leptailurus, Mayailurus, Otocolobus, Prionailurus, Profelis, Puma*); small cats; Americas, Eurasia, Africa; mainly forest.

F. aurata	African golden cat	Senegal – Zaire – Kenya; forest
F. badia	Bay cat (Bornean red cat)	Borneo; R
F. bengalensis	Leopard cat	E Siberia – Baluchistan – Java, Borneo, Taiwan, W Philippines
F. bieti	Chinese desert cat	W China, S Mongolia; steppe
F. caracal	Caracal	Turkestan, NW India – Arabia; Africa; steppe, savanna; (R)
F. chaus	Jungle cat	Egypt – Indochina, Sri Lanka; dry forest
F. colocolo	Pampas cat	Ecuador – Patagonia; grassland, forest
F. concolor	Puma (Cougar)	S Canada – Patagonia; forest – steppe; E
F. geoffroyi	Geoffroy's cat	Bolivia – Patagonia
F. guigna	Kodkod	C, S Chile, W Argentina; forest
F. iriomotensis	Iriomote cat	Iriomote I, Ryukyu Is; close to *F. bengalensis*; E
F. jacobita	Andean cat	S Peru – N Chile; montane steppe; R
F. manul	Pallas's cat	Iran – W China; montane steppe
F. margarita	Sand cat	Sahara – Baluchistan; desert; (E)
F. marmorata	Marbled cat	Nepal – Malaya, Sumatra, Borneo; I
F. nigripes	Black-footed cat (Small spotted cat)	S Africa, Botswana, Namibia; steppe, savanna
F. pardalis	Ocelot	Arizona – N Argentina; forest, scrub; V
F. planiceps	Flat-headed cat	Malaya, Sumatra, Borneo; I
F. rubiginosa	Rusty-spotted cat	S India, Sri Lanka; K
F. serval	Serval	Africa; savanna
F. silvestris (*catus*) (*lybica*)	Wild cat	W Europe – India; Africa; open forest, savanna, steppe; ancestor of domestic cat *F. catus*

F. temminckii	Asiatic golden cat	Nepal – S China – Sumatra; forest; I
F. tigrina	Littled spotted cat (Oncilla)	Costa Rica – N Argentina; forest; V
F. viverrina	Fishing cat	India – S China – Java; forest, swamps
F. wiedii	Margay	N Mexico – N Argentina; forest; V
F. yagouaroundi	Jaguarundi	Arizona – C Argentina; forest, scrub; I

Lynx; (*Felis*); lynxes; N America, N Eurasia.

L. canadensis (*lynx*)	Canadian lynx	Canada, Alaska, N USA; con. forest
L. lynx (*pardina*)	Eurasian lynx	W Europe – Siberia; con. forest; (E)
L. rufus	Bobcat	S Canada – S Mexico

Panthera; (*Leo, Tigris, Uncia*); big cats; Americas, Asia, Africa; mainly forest.

P. leo	Lion	Africa S of Sahara; NW India; († NW Africa, SW Asia); savanna, steppe; (E)
P. onca	Jaguar	N Mexico – N Argentina; († S USA, Uruguay, C Argentina); forest; V
P. pardus	Leopard (Panther)	SE Siberia – Java – Asia Minor; Africa; forest; T
P. tigris	Tiger	SE Siberia – Java – Caucasus; forest; E
P. uncia	Snow leopard (Ounce)	Altai – Himalayas; montane steppe; E

Tiger
(Panthera tigris)

Neofelis; (*Panthera*).

| *N. nebulosa* | Clouded leopard | Nepal – S China – Sumatra, Borneo, Taiwan; forest; V |

Acinonyx

| *A. jubatus* | Cheetah | Baluchistan – Arabia; Africa; steppe, savanna; E/V |

ORDER PINNIPEDIA

Seals, Sealions etc.; c. 34 species; worldwide except tropical W Pacific and Indian Ocean; marine, (freshwater); breed on shore or ice; mainly predators on fish; sometimes included in order Carnivora; refs. 7.1, 2, 3.

Family Otariidae

Eared seals, sealions; c. 14 species; S Hemisphere, N Pacific; refs. 7.1, 2.

Northern sealion
(Eumetopias jubatus)

Arctocephalus; (*Arctophoca*); southern fur seals; S Hemisphere – California.

A. australis	South American fur seal	Brazil – Peru
A. forsteri	New Zealand fur seal	S New Zealand, etc., S Australia
A. galapagoensis	Galapagos fur seal	Galapagos Is
A. gazella	Antarctic fur seal	Islands S of Antarctic Convergence
A. philippii	Juan Fernandez fur seal	Juan Fernandez Is; V
A. pusillus (*doriferus*) (*tasmanicus*)	Afro-Australian fur seal (Cape fur seal)	S Africa, S Australia, Tasmania
A. townsendi	Guadalupe fur seal	S California – W Mexico; V
A. tropicalis	Subantarctic fur seal (Amsterdam Island fur seal)	Islands N of Antarctic Convergence

Callorhinus

C. ursinus	Northern fur seal	Bering, Okhotsk Seas

Zalophus

Z. californianus	Californian sealion	California, Galapagos, († Japan); (E)

Eumetopias

E. jubatus	Northern sealion (Steller's sealion)	N Japan – California

Otaria

O. byronia (*flavescens*)	South American sealion (Southern sealion)	S Brazil – Peru

Neophoca

N. cinerea	Australian sealion	SW, S Australia

Phocarctos; (*Neophoca*).

P. hookeri	New Zealand sealion (Hooker's sealion)	S of New Zealand

Family Odobenidae

One species.

Odobenus

O. rosmarus	Walrus	Arctic ocean; (K)

*Walrus
(Odobenus rosmarus)*

Family Phocidae

Earless seals; 19 species; polar, temperate, some tropical seas; formerly in order Pinnipedia but probably more closely related to the Mustelidae than to the Otariidae; refs. 7.1, 2.

*Hooded seal
(Cystophora cristata)*

Phoca; (*Pusa, Histriophoca, Pagophilus*); N temperate and polar seas.

P. caspica	Caspian seal	Caspian Sea
P. fasciata	Ribbon seal (Banded seal)	N Pacific; subarctic
P. groenlandicus	Harp seal (Greenland seal)	N Atlantic; subarctic; breed on ice
P. hispida	Ringed seal	Arctic Ocean, N Pacific, Baltic; breed on ice; (E)
P. largha	Larga seal	NW Pacific; breed on ice
P. sibirica	Baikal seal	Lake Baikal
P. vitulina (*kurilensis*) (*richardii*)	Common seal (Harbour seal) (Spotted seal)	N Atlantic, N Pacific; temperate, subarctic; (V)

Halichoerus

H. grypus	Grey seal (Atlantic seal)	Newfoundland, etc., Britain – White Sea, Baltic

Erignathus

E. barbatus	Bearded seal	Arctic Ocean; breed on ice

Lobodon

L. carcinophagus	Crabeater seal	Antarctic; edge of pack ice

Ommatophoca

O. rossi	Ross seal	Antarctic; edge of pack ice

Hydrurga

H. leptonyx	Leopard-seal	Southern Ocean

Leptonychotes

L. weddelli	Weddell seal	Antarctic; edge of ice

Monachus; monk seals; northern tropical and warm temperate zones.

M. monachus	Mediterranean monk seal	Canary Is – Mediterranean, Black Sea; E

M. schauinslandi	Hawaiian monk seal	Hawaii; E
M. tropicalis †	Caribbean monk seal	Formerly Caribbean: probably extinct

Mirounga; elephant-seals.

M. angustirostris	Northern elephant-seal	W coast of N America
M. leonina	Southern elephant-seal	Subantarctic Islands, Patagonia

Cystophora

C. cristata	Hooded seal (Bladder-nose)	N Atlantic; edge of ice

ORDER CETACEA

Whales, dolphins, porpoises; *c.* 77 species; all oceans, some tropical rivers; ref. 7.4, 5.

SUBORDER ODONTOCETI

Toothed whales; *c.* 65 species; all oceans, some tropical rivers.

Family Platanistidae

La plata dolphin (Pontoporia blainvillei)

River dolphins; *c.* 4 species; S America, S, E Asia; rivers, coastal waters; fish-eaters.

Pontoporia

P. blainvillei	La Plata dolphin (Franciscana)	S Brazil – N Argentina; coasts, estuaries; K

Inia

I. geoffrensis	Boutu (Amazon porpoise)	Amazon, upper Orinoco; V

Lipotes

L. vexillifer	White fin dolphin	Yangtzekiang R, Tungting L, China; E

Platanista; (*Susu*).

P. gangetica (*minor*) (*indi*)	Ganges dolphin	Ganges, Brahmaputra, Indus & tributaries; E, V

Family Delphinidae

Common dolphin (Delphinus delphis)

Marine dolphins; *c.* 34 species; all oceans, some tropical rivers; mainly fish-eaters.

Steno

S. bredanensis	Rough-toothed dolphin	Tropical, warm temperate seas

Sotalia

S. fluviatilis	Tucuxi	Panama – S Brazil;
(*brasiliensis*)		Amazon; coasts, rivers
(*guianensis*)		

Sousa; hump-backed dolphins; tropical coasts, estuaries, large rivers.

S. chinensis	Indo-Pacific hump-backed	Indian Ocean, SW Pacific,
(*borneensis*)	dolphin	lower Yangtzekiang
(*lentiginosus*)		
(*plumbea*)		
S. teuszii	Atlantic hump-backed	Mauritania – Angola
	dolphin	

Stenella; tropical and warm temperate seas; classification provisional.

S. attenuata	Pantropical spotted	Tropical oceans
(*dubia*)	dolphin	
(*graffmani*)	(Bridled dolphin)	
S. clymene	Atlantic spinner dolphin	Atlantic
S. coeruleoalba	Striped dolphin	Atlantic, Pacific; tropical,
(*styx*)	(Euphrosyne dolphin)	temperate
S. frontalis	Atlantic spotted dolphin	Atlantic; tropical, warm
(*plagiodon*)		temperate; ref. 7.6
S. longirostris	Spinner dolphin	Tropical, warm temperate
	(Long-beaked dolphin)	seas

Delphinus

D. delphis	Common dolphin	Temperate, tropical seas;
(*bairdii*)		Black Sea
D. tropicalis	Tropical dolphin	Indian Ocean, etc., tropical

Tursiops

T. truncatus	Bottle-nosed dolphin	Atlantic, Indian Oc.,
(*gilli*)		Pacific; tropical,
		temperate

Lissodelphis; right whale dolphins; Pacific, S Atlantic.

L. borealis	Northern right whale	N Pacific; temperate
	dolphin	
L. peronii	Southern right whale	S Atlantic, S Pacific;
	dolphin	temperate, subtropical

Lagenodelphis

L. hosei	Fraser's dolphin	Pacific, Indian Oc.,
		W Atlantic; tropical

Lagenorhynchus; short-beaked dolphins; all temperate and polar seas.

L. acutus	Atlantic white-sided	N Atlantic, Atlantic
	dolphin	Arctic; cold temperate, polar
L. albirostris	White-beaked dolphin	N Atlantic, Atlantic
		Arctic; cold temperate, polar

L. australis	Peale's dolphin	Southern S America; coastal
L. cruciger	Hourglass dolphin	S Atlantic, S Pacific, Southern Oc.; temperate
L. obliquidens	Pacific white-sided dolphin	N Pacific; temperate
L. obscurus	Dusky dolphin	Southern Oc.; inshore

Peponocephala; (*Lagenorhynchus*).

P. electra	Melon-headed whale	Tropical seas

Cephalorhynchus; piebald dolphins; southern temperate seas; coastal.

C. commersoni	Commerson's dolphin	Patagonia, S Georgia, Kerguelen; K
C. eutropia	Black dolphin	Chile; K
C. heavisidii	Heaviside's dolphin	S Africa; K
C. hectori	Hector's dolphin	New Zealand; K

Orcaella

O. brevirostris	Irrawaddy dolphin	Bay of Bengal – N Australia; coastal, large rivers; K

Pseudorca

P. crassidens	False killer whale	All seas; tropical, temperate

Orcinus

O. glacialis	Southern killer	Southern Ocean (Indian Ocean sector); ref. 7.7
O. orca	Killer whale	All seas; mainly polar, temperate

Grampus; (*Grampidelphis*).

G. griseus (*rectipinna*)	Risso's dolphin	All seas; mainly temperate

Globicephala; pilot whales, blackfish.

G. macrorhynchus (*scammonii ?*) (*melas*)	Short-finned pilot whale	Subtropical Atlantic, Pacific
G. melas (*edwardi*) (*melaena*)	Long-finned pilot whale	N Atlantic, southern Oc.; temperate

Feresa

F. attenuata (*occulta*)	Pygmy killer whale (Slender blackfish)	Atlantic, Pacific; tropical, warm temperate

Family Phocoenidae

Porpoises; *c.* 6 species; S temperate, N temperate and Arctic coasts; mainly fisheaters.

*Common porpoise
(Phocoena phocoena)*

Phocoena

P. dioptrica	Spectacled porpoise	La Plata – S Georgia, ? New Zealand
P. phocoena (*vomerina*)	Common porpoise (Harbour porpoise)	N Atlantic, N Pacific, Arctic Oc., Black Sea; K
P. sinus	Gulf porpoise (Cochito)	Gulf of California; V
P. spinipinnis	Burmeister's porpoise	La Plata – Peru

Phocoenoides

P. dalli (*truei*)	Dall's porpoise	N Pacific; temperate

Neophocaena

N. phocaenoides (*asiaeorientalis*) (*sunameri*)	Finless porpoise	Iran – Borneo – Japan; Yangtzekiang; mainly coastal

Family Monodontidae

White whales; 2 species; Arctic Ocean.

*Narwhal
(Monodon monoceros)*

Delphinapterus

D. leucas	White whale (Beluga)	Arctic Ocean; K

Monodon

M. monoceros	Narwhal	Arctic Ocean; K

Family Physeteridae

Sperm whales; 3 species; all oceans; squid-eaters.

*Sperm whale
(Physeter macrocephalus)*

Kogia

K. breviceps	Pygmy sperm whale	Tropical, warm temperate seas
K. simus	Dwarf sperm whale	Tropical, subtropical

Physeter

P. macrocephalus (*catodon*)	Sperm whale	All oceans; tropical, (temperate)

Family Ziphiidae

Beaked whales; *c.* 18 species; all oceans; mainly squid-eaters.

*Cuvier's beaked whale
(Ziphius cavirostris)*

Tasmacetus

T. shepherdi	Shepherd's beaked whale (Tasman whale)	New Zealand, S America

Berardius

B. arnuxii	Arnoux's beaked whale (Southern four-toothed whale)	Southern hemisphere; temperate
B. bairdii	Baird's beaked whale (Northern four-toothed whale)	N Pacific; temperate

Indopacetus; (*Mesoplodon*); known only from two skulls.

I. pacificus	Indo-pacific beaked whale	Queensland; Somalia

Mesoplodon; (*Micropteron*).

M. bidens	Sowerby's beaked whale	N Atlantic, Baltic; temperate
M. bowdoini	Andrew's beaked whale	SW Pacific, Indian Oc.
M. carlhubbsi	Hubb's beaked whale	N Pacific; warm temperate
M. densirostris	Blainville's beaked whale	All tropical & warm temperate seas
M. europaeus (*gervaisi*)	Gervais' beaked whale	N Atlantic, mainly western; subtropical, warm temperate
M. ginkgodens	Ginkgo-toothed beaked whale	N Pacific, Indian Oc.
M. grayi	Gray's beaked whale	S Pacific, S Indian Oc., S Atlantic, (NE Atlantic); temperate
M. hectori	Hector's beaked whale	S Hemisphere, N Pacific; temperate seas
M. layardii	Strap-toothed whale (Layard's beaked whale)	S Hemisphere; temperate seas
M. mirus	True's beaked whale	N Atlantic; S Africa; temperate
M. stejnegeri	Stejneger's beaked whale	N Pacific; temperate

Ziphius

Z. cavirostris	Cuvier's beaked whale (Goose-beaked whale)	All oceans; temperate, tropical

Hyperoodon; bottle-nose whales.

H. ampullatus	Northern bottlenose whale	Arctic Oc., N Atlantic; V
H. planifrons	Southern bottlenose whale	S Atlantic, Indian Oc., S Pacific, Southern Oc.

SUBORDER MYSTICETI

Baleen whales; 10 species; all oceans; filter-feeders.

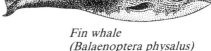

Grey whale
(*Eschrichtius robustus*)

Family Eschrichtiidae

One species.

Eschrichtius; (*Rhachianectes*).

E. robustus (*glaucus*) (*gibbosus*)	Grey whale	NE, (NW) Pacific; coastal († N Atlantic)

Family Balaenopteridae

Rorquals; 6 species; all oceans.

Balaenoptera; (*Sibbaldus*); rorquals; all oceans; off-shore.

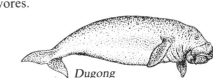

Fin whale
(*Balaenoptera physalus*)

B. acutorostrata	Minke whale (Lesser rorqual)	All oceans
B. borealis	Sei whale	All oceans
B. edeni	Bryde's whale	All oceans
B. musculus (*brevicauda*)	Blue whale	All oceans; E
B. physalus	Fin whale (Common rorqual)	All oceans; V

Megaptera

M. novaeangliae	Humpback whale	All oceans; E

Family Balaenidae

Right whales; 3 species; temperate, polar seas; mainly coastal.

Caperea

C. marginata	Pygmy right whale	S Hemisphere; temperate, polar

Black right whale
(*Balaena glacialis*)

Balaena; (*Eubalaena*).

B. glacialis (*australis*)	Black right whale	All temperate and subarctic seas; E/V
B. mysticetus	Bowhead (Greenland right whale)	Arctic Ocean; E

ORDER SIRENIA

Sea cows; 5 species (one of them extinct); coasts of Indian Ocean, coasts and adjacent rivers of tropical Atlantic Ocean (formerly also N Pacific); herbivores.

Family Dugongidae

Dugong

D. dugon	Dugong	E Africa, Red Sea – N Australia; coastal; V

Dugong
(*Dugong dugon*)

Hydrodamalis

H. gigas †	Steller's sea cow	Bering Sea, N Pacific; extinct

Family Trichechidae

Manatees; 3 species; coasts of tropical Atlantic and adjacent rivers.

Caribbean Manatee (Trichechus)

Trichechus

T. inunguis	Amazon manatee	R Amazon; V
T. manatus	Caribbean manatee	Virginia – W Indies – Brazil; coastal; V
T. senegalensis	African manatee	Senegal – Angola; rivers; V

ORDER PROBOSCIDEA

Family Elephantidae

Elephants; 2 species; Africa, SE Asia; forest, savanna; herbivores.

African Elephant (Loxodonte africana)

Loxodonta

L. africana (*cyclotis*)	African elephant	Africa S of Sahara; forest, savanna; V

Elephas

E. maximus	Indian elephant	India – Sumatra, Sri Lanka; [Borneo]; forest; V

ORDER PERISSODACTYLA

Odd-toed ungulates; 16 species; Eurasia, Africa, C, S America; herbivores.

Family Equidae

Horses etc.; 7 species; Africa, W, C Asia; steppe, savanna; grazers. Classification of the Asiatic asses remains unstable.

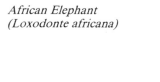

Common zebra (Equus burchellii)

Equus; (*Hemionus, Hippotigris*).

E. africanus (*asinus*)	African ass	NE Africa; ancestor of domestic ass *E. asinus*; E
E. burchellii (*quagga*)	Common zebra (Burchell's zebra)	E, S Africa; steppe, savanna
E. ferus (*caballus*) (*przewalskii*)	Wild horse	Mongolia, Sinkiang; († E Europe); ancestor of domestic horse *E. caballus*; Ex ?
E. grevyi	Grevy's zebra	Kenya, Somalia, S Ethiopia; steppe; E

E. hemionus
  (*kiang*)
  (*onager*)
	Asiatic wild ass
	  (Kulan)
	  (Kiang)
	  (Onager)
		Iran – Pakistan, Tibet,
		  Mongolia (†Asia Minor,
		  S Russia, etc.); steppe,
		  subdesert; ref. 8.8; E/V

E. quagga †	Quagga	S Africa; extinct
E. zebra	Mountain zebra	SW Africa; E/V

Family Tapiridae

Tapirs; 4 species; C, S America, SE Asia; forest; browsers.

Tapirus; (*Acrocodia*).

T. bairdii	Baird's tapir	S Mexico – Ecuador; V
T. indicus	Malayan tapir	Burma – Sumatra; E

T. pinchaque
  (*roulini*)
	Mountain tapir
	  (Woolly tapir)
		Colombia, Ecuador; V

T. terrestris	Brazilian tapir	Colombia – S Brazil

Brazilian tapir
(Tapirus terrestris)

Family Rhinocerotidae

Rhinoceroses; 5 species; Africa, SE Asia; forest, savanna; browsers, grazers.

Rhinoceros; SE Asia; high grass.

R. sondaicus	Javan rhinoceros	Java († Malaya – India); E
R. unicornis	Indian rhinoceros	Nepal, NE India; E

Dicerorhinus; (*Didermoceros*).

D. sumatrensis	Sumatran rhinoceros	Borneo to Malaya,
		  (formerly to Assam);
		  forest; E

Javan rhinocerus
(Rhinoceros sondaicus)

Ceratotherium; (*Diceros*).

C. simum	White rhinoceros
	  (Square-lipped
	  rhinoceros)
		Uganda etc., Zululand;
		  savanna; grazer; (E)

Diceros

D. bicornis	Black rhinoceros	Africa S of Sahara; savanna; browser; E

ORDER HYRACOIDEA

Family Procaviidae

Hyraxes (dassies); *c.* 8 species; Africa, Arabia; forest, rock outcrops in savanna and steppe; herbivores.

Dendrohyrax; tree hyraxes; forest.

D. arboreus	Kenya – S Africa
D. dorsalis	Gambia – Uganda
D. validus	E Tanzania, Zanzibar, etc.

Rock hyrax
(Procavia capensis)

Heterohyrax

H. brucei	Small-toothed rock hyrax	SE Egypt – Transvaal

Procavia; large-toothed rock hyraxes; classification very provisional.

P. capensis (*johnstoni*)	S Africa – S Malawi
P. habessinica	Ethiopian Highlands
P. syriaca (*ruficeps*)	Senegal – N Tanzania – Egypt, Syria, Arabia
P. welwitschii	N Namibia, S Angola

Aardvark
(Orycteropus afer)

ORDER TUBULIDENTATA

Family Orycteropidae

One species.

Orycteropus

O. afer	Aardvark (Antbear)	Africa S of Sahara; savanna, steppe

ORDER ARTIODACTYLA

Even-toed ungulates; *c.* 194 species; Americas, Eurasia, Africa; herbivores.

Family Suidae

Pigs; 8 species; Eurasia, Africa; forest, savanna; omnivores.

Wart hog
(Phacochoerus aethiopicus)

Potamochoerus

P. porcus	Bush pig (Red river hog)	Africa S of Sahara, Madagascar; forest, savanna

Sus; Eurasian pigs; Eurasia, (NW Africa); forest.

S. barbatus	Bearded pig	Malaya, Sumatra, Borneo, Philippines etc.; V
S. salvanius	Pygmy hog	SE Nepal, etc.; E
S. scrofa	Wild boar	Europe, NW Africa – SE Siberia – Java, Honshu, Taiwan, Sri Lanka, [New Guinea, New Zealand]; ancestor of domestic pig, *S. domesticus*; V
S. verrucosus (*celebensis*)	Javan pig	Java, Sulawesi, Philippines; V

Phacochoerus

P. aethiopicus	Wart hog	Africa S of Sahara; steppe, savanna

Hylochoerus

H. meinertzhageni	Giant forest hog	Liberia – SW Ethiopia; forest

Babyrousa
B. babyrussa   Babirusa   Sulawesi, etc.; forest; V

Family Tayassuidae

Peccaries; 3 species; Texas – S America; desert – forest.

Tayassu; (*Dicotyles*).
T. pecari   White-lipped peccary   Mexico – N Argentina;
  (*albirostris*)   [Cuba]; forest
T. tajacu   Collared peccary   S USA – C Argentina;
                                                  [Cuba]; desert –
                                                  forest

Coillared peccary
(Tayassu tajacu)

Catagonus
C. wagneri   Chaco peccary   Paraguay, N Argentina,
                                                  Bolivia; grassland,
                                                  scrub; ref. 8.1; V

Family Hippopotamidae

Hippopotamuses; 2 species; Africa; grazers.

Hippopotamus
H. amphibius   Hippopotamus   Africa S of Sahara;
                                                  grassland with rivers or
                                                  lakes

Hippopotamus
(Hippopotamus amphibius)

Choeropsis
C. liberiensis   Pygmy hippopotamus   Sierra Leone – Nigeria;
                                                  forest by fresh water; V

Family Camelidae

Camels, llamas; 4 species; S America, SW, C Asia, ? N Africa; desert, steppe.

Lama
L. guanicoe   Guanaco   Peru – Patagonia; possible
                                                  ancestor of domestic
                                                  llama, *L. glama* and
                                                  alpaca, *L. pacos*

Dromedary
(Camelus dromedarius)

Vicugna; (*Lama*).
V. vicugna   Vicugna   Peru – N Chile;
                                                  montane; V

Camelus
C. ferus   Bactrian camel   C Asia; ancestor of
  (*bactrianus*)   (Two-humped camel)   domestic Bactrian camel
                                                  C. bactrianus; V
C. dromedarius   Dromedary   ? Arabia, only known as
                       (Arabian camel)   domestic animal
                       (One-humped camel)

Family Tragulidae

Chevrotains (mouse-deer); *c.* 4 species; W, C Africa, SE Asia; forest.

*Water chevrotain
(Hyemoschus aquaticus)*

Hyemoschus

H. aquaticus	Water chevrotain	Sierra Leone – W Uganda

Tragulus

T. javanicus	Lesser Malay chevrotain	Indochina – Java, Borneo
T. meminna	Indian spotted chevrotain	India, Sri Lanka
T. napu	Greater Malay chevrotain	S Indochina – Sumatra, Borneo

Family Moschidae

Musk deer; 5 species; E Asia; forest; classification very unstable.

*Siberian musk deer
(Moschus moschiferus)*

Moschus

M. berezovskii		SW China, etc.
M. chrysogaster	Forest musk deer	W Himalayas – Gansu; V
M. fuscus		Yunnan, China; ref. 8.2
M. moschiferus (*sibiricus*)	Siberian musk deer	E Siberia
M. sifanicus	Alpine musk deer	W Himalayas – Shaanxi

Family Cervidae

Deer; *c.* 38 species; Americas, Eurasia, (NW Africa); mainly forest; browsers, grazers.

*Chinese muntjac
(Muntiacus reevesi)*

Hydropotes

H. inermis	Chinese water deer	EC China, Korea, [England]

Muntiacus; muntjacs (barking deer); E Asia.

M. atherodes (*pleiharicus*)	Bornean yellow muntjac	Borneo; ref. 8.3
M. crinifrons	Black muntjac	Zhejiang, etc., E China; I
M. feae	Fea's muntjac	Thailand; E
M. muntjak	Indian muntjac	India – S China – Java, Borneo, Sri Lanka, Taiwan
M. reevesi	Chinese muntjac	S, E China, Taiwan, [England, France]
M. rooseveltorum	Roosevelts' muntjac	Indochina

Elaphodus

E. cephalophus	Tufted deer	S China, N Burma

Cervus; (*Axis, Dama, Sika*); Eurasia, (N America).

C. albirostris	Thorold's deer	E Tibetan Plateau; I

C. alfredi		Negros, Panay etc., Philippines; ref. 8.4; E
C. axis	Spotted deer (Chital)	India etc., Sri Lanka, [New Zealand, Argentina, USA, etc.]
C. dama (*mesopotamica*)	Fallow deer	S, [W, C] Europe – S Iran, [N, S America, S Africa, Australia, New Zealand, etc.]; (E)
C. duvaucelii	Swamp deer (Barasingha)	India, SW Nepal; E
C. elaphus (*canadensis*)	Red deer, Wapiti (American elk)	W Europe, NW Africa – W China; W Canada, W USA; [New Zealand, Argentina]; (E/V)
C. eldii	Thamin	Assam – Indochina, Hainan; E
C. mariannus (*unicolor*)	Philippine rusa	Marianas Is, Philippines; ref. 8.4
C. nippon	Sika deer	SE Siberia – E China, Japan, Taiwan, [W Europe, New Zealand]; (E)
C. porcinus (*calamianensis*) (*kuhli*)	Hog-deer	NW India – Indochina, Bawean I (Java), Calamian Is (Philippines), [Australia, Sri Lanka]; (V/R)
C. schomburgki †	Schomburgk's deer	Thailand; extinct
C. timorensis	Timor deer (Rusa deer)	Java, Sulawesi, Timor, etc., [N Australia, New Zealand]
C. unicolor	Sambar	India – S China – Java, Borneo, Sulawesi, Philippines, Sri Lanka, [Australia, New Zealand]
Elaphurus		
E. davidianus	Père David's deer	NE China; extinct in wild
Alces		
A. alces	Moose, Elk (Europe)	Scandinavia – E Siberia; Alaska, Canada, N USA; [New Zealand]; con. forest
Rangifer		
R. tarandus (*caribou*)	Reindeer, Caribou	Scandinavia – E Siberia; Alaska, Canada, Greenland; tundra; includes domestic reindeer

Odocoileus; (*Blastocerus, Ozotoceros*).

O. bezoarticus (*campestris*)	Pampas deer	E Brazil – N Argentina; grassland; (E)
O. hemionus	Mule deer (Black-tailed deer)	S Alaska – W, C USA – N Mexico; (R)
O. virginianus	White-tailed deer	S Canada – Peru, N Brazil; [Cuba, etc., New Zealand]; (R)

Blastocerus; (*Odocoileus*).

B. dichotomus	Marsh deer	S Brazil – NE Argentina; V

Hippocamelus; guemals (huemuls); Andes.

H. antisensis	Peruvian guemal	Ecuador – NW Argentina; V
H. bisulcus	Chilean guemal	S, C Chile, SW Argentina; E

Mazama; brockets; C, S America.

M. americana	Red brocket	C Mexico – N Argentina
M. chunyi	Dwarf brocket	S Peru, N Bolivia
M. gouazoubira	Brown brocket	S Mexico – N Argentina
M. rufina	Little red brocket	Venezuela – Ecuador

Pudu; pudus; Andes.

P. mephistophiles	Northern pudu	Colombia – N Peru; I
P. pudu	Southern pudu	S Chile, SW Argentina

Capreolus; roe deer; ref. 8.9

C. capreolus	Western roe deer	W Europe – W Siberia
C. pygargus	Eastern roe deer	Tien Shan, C Siberia – S China

Family Giraffidae

Giraffe, Okapi; 2 species, Africa; browsers.

Okapia

O. johnstoni	Okapi	Zaire; forest

Giraffa

G. camelopardalis	Giraffe	Africa S of Sahara; savanna

Giraffe
(Giraffa camelopardalis)

Family Antilocapridae

One species; sometimes included in the Bovidae.

Antilocapra

A. americana	Pronghorn	W, C USA – N Mexico; [Argentina]; desert, grassland; (E)

Pronghorn
(Antilocapra americana)

Family Bovidae

Cattle, antelope, sheep, goats; *c.* 127 species; N America, Eurasia, Africa; grazers, browsers.

Tragelaphus; (*Limnotragus, Boocercus, Strepsiceros, Taurotragus*).

T. angasii	Nyala	Natal – Malawi; savanna
T. buxtoni	Mountain nyala	Ethiopia; montane forest, grassland; E
T. derbianus	Giant eland	Senegal – Nile; savanna; (E)
T. eurycerus	Bongo	Sierra Leone – Kenya; forest
T. imberbis	Lesser kudu	Ethiopia – Tanzania, ? S Arabia; steppe
T. oryx	Eland	E, S Africa; savanna
T. scriptus	Bushbuck	Africa S of Sahara; forest, scrub
T. spekii	Sitatunga	Africa S of Sahara; swamps
T. strepsiceros	Greater kudu	Chad – Ethiopia – S Africa; savanna

Greater kudu
(Tragelaphus strepsiceros)

Boselaphus

B. tragocamelus	Nilgai	India; forest, scrub

Tetracerus

T. quadricornis	Four-horned antelope	India; woodland

Bubalus; (*Anoa*); Asiatic buffaloes; SE Asia.

B. arnee (*bubalis*)	Water buffalo	India; ancestor of domestic *B. bubalis*; E
B. depressicornis	Lowland anoa	Sulawesi; forest; E
B. mindorensis	Tamarau	Mindoro, Philippines; swamp, forest; E
B. quarlesi	Mountain anoa	Sulawesi; E

Bos; (*Bibos*); oxen; Eurasia; forest.

B. gaurus	Gaur (Indian bison)	India – Malaya; forest; ancestor of domestic mithan (gayal), *B. frontalis*; V
B. javanicus (*banteng*)	Banteng	Burma – Java, Borneo; forest; includes domestic Bali cattle; V
B. mutus (*grunniens*)	Yak	Tibetan Plateau; montane grassland; ancestor of domestic yak, *B. grunniens*; E

B. primigenius (*taurus*)	Aurochs (Urus)	Europe, W Asia; extinct in wild but ancestor of domestic cattle, *B. taurus*, including zebu, *B. indicus*
B. sauveli	Kouprey	Indochina; forest; E

Synceros

S. caffer (*nanus*)	African buffalo	Africa S of Sahara; forest, savanna

Bison; ref. 8.10

Bison
(Bison bison)

B. bison (*bonasus*)	Bison (Wisent)	N America, E Europe; grassland (woodland); (V)

Cephalophus; forest duikers; Africa S of Sahara; forest.

C. adersi	Aders' duiker	Zanzibar, E Kenya; V
C. callipygus	Peters' duiker	Gabon, Cameroun, etc.
C. dorsalis	Bay duiker	Guinea – Zaire, etc.
C. jentinki	Jentink's duiker	Liberia, etc.; E
C. leucogaster	White-bellied duiker	Cameroun – E Zaire
C. natalensis	Red forest duiker	Somalia – E Zaire – S Africa
C. niger	Black duiker	Guinea – Nigeria
C. nigrifrons	Black-fronted duiker	Cameroun – Kenya – Angola; wet forest
C. ogilbyi	Ogilby's duiker	Sierra Leone – Gabon
C. rufilatus	Red-flanked duiker	Senegal – Sudan; forest-edge, thicket
C. spadix	Abbott's duiker	Tanzania; montane; V
C. sylvicultor	Yellow-backed duiker	Gambia – Kenya – Angola
C. weynsi		Zaire – W Kenya
C. zebra	Banded duiker	Liberia, etc.; V

Philantomba; (*Cephalophus*).

P. maxwellii	Maxwell's duiker	Senegal – Nigeria
P. monticola	Blue duiker	SE Nigeria – Kenya – S Africa

Sylvicapra

S. grimmia	Common duiker	Africa S of Sahara; savanna

Kobus; (*Adenota, Onotragus*); Africa S of Sahara.

K. ellipsiprymnus (*defassa*)	Waterbuck	Africa S of Sahara; savanna
K. kob	Kob	Volta – Kenya; savanna
K. leche	Lechwe	Zambia, etc.; wet grassland; V

| *K. megaceros* | Nile lechwe | S Sudan; swamps |
| *K. vardonii* | Puku | Zambia, etc.; savanna |

Redunca; reedbuck; Africa S of Sahara; savanna.

R. arundinum	Reedbuck	S Africa – L Victoria
R. fulvorufula	Mountain reedbuck	S, E, C Africa
R. redunca	Bohar reedbuck	Senegal – Tanzania

Pelea

| *P. capreolus* | Grey rhebok | S Africa; hilly grassland |

Hippotragus

H. equinus	Roan antelope	Africa S of Sahara; savanna
H. leucophaeus †	Blue buck	SW Africa; extinct
H. niger	Sable antelope	S Africa – Kenya; savanna; (E)

Oryx

O. dammah (*tao*)	Scimitar oryx	S edge of Sahara; semi-desert; E
O. gazella (*beisa*)	Gemsbok (Beisa)	SW, E Africa; steppe
O. leucoryx	Arabian oryx	Arabia; desert; E

Addax

| *A. nasomaculatus* | Addax | Sahara; desert, semi-desert; E |

Connochaetes; (*Gorgon*); wildebeest; steppe.

| *C. gnou* | Black wildebeest (White-tailed gnu) | S Africa |
| *C. taurinus* | Blue wildebeest (Brindled gnu) | S, E Africa |

Alcelaphus

| *A. buselaphus* | Red hartebeest | W, E, SW Africa; steppe; (E) |

Sigmoceros

| *S. lichtensteinii* | Lichtenstein's hartebeest | Angola – Tanzania; savanna |

Damaliscus; (*Beatragus*).

D. dorcas	Bontebok, Blesbok	S Africa; steppe; (V)
D. hunteri	Hunter's hartebeest	NE Kenya, S Somalia; steppe; R
D. lunatus (*korrigum*)	Tsessebi, Topi	Africa S of Sahara; savanna

Oreotragus
O. oreotragus	Klipspringer	Africa S of Sahara; rocky hills in savanna

Madoqua; (*Rhynchotragus*); dik-diks; E, S Africa; dry steppe; ref. 8.5.
M. guentheri	Gunther's dik-dik	E Africa
M. kirkii	Kirk's dik-dik (Damara dik-dik)	SW, E Africa
M. piacentinii		SE Somalia
M. saltiana (*phillipsi*) (*swaynei*)	Salt's dik-dik (Swayne's dik-dik)	N, E Ethiopia, Somalia

Dorcatragus
D. megalotis	Beira antelope	Somalia, E Ethiopia; semi-desert; K

Ourebia
O. ourebi	Oribi	Africa S of Sahara; steppe, savanna

Raphiceros
R. campestris	Steenbok	S, E Africa; steppe, savanna
R. melanotis	Cape grysbok	SW Africa; steppe
R. sharpei	Sharpe's grysbok	S Africa – Tanzania; savanna

Neotragus; (*Nesotragus*).
N. batesi	Bates' dwarf antelope	SE Nigeria – E Zaire, etc.; forest
N. moschatus	Suni	Kenya – Natal; thicket; (E)
N. pygmaeus	Royal antelope	Sierra Leone – Ghana; forest

Aepyceros
A. melampus	Impala	S, E Africa; savanna; (E)

Antilope
A. cervicapra	Blackbuck	India; [Texas]; steppe

Antidorcas
A. marsupialis	Springbok	S Africa; steppe

Litocranius
L. walleri	Gerenuk	E Africa; steppe

Ammodorcas
A. clarkei	Dibatag	Somalia, E Ethiopia; semi-desert; V

Gazella; gazelles; Sahara – Tanzania – Mongolia; desert, steppe.

G. arabica		Farsan I, Red Sea
G. bennettii (*dorcas*)	Indian gazelle	Iran – C India; ref. 8.11
G. bilkis		Yemen; ref. 8.7
G. cuvieri	Edmi gazelle	Morocco, N Algeria, Tunis; E
G. dama	Addra gazelle	W, S Sahara; V
G. dorcas	Dorcas gazelle	Sahara – Arabia; V
G. gazella	Mountain gazelle (Idmi)	Arabia – Israel; V
G. granti	Grant's gazelle	Ethiopia – Tanzania
G. leptoceros	Sand gazelle (Rhim)	N Sahara; V
G. rufifrons	Red-fronted gazelle	Senegal – Ethiopia; V
G. rufina †	Red gazelle	Algeria; extinct
G. soemmerringii	Soemmerring's gazelle	Sudan, Ethiopia, Somalia; V
G. spekei	Speke's gazelle (Dero)	Somalia, E Ethiopia; V
G. subgutturosa	Goitred gazelle	Arabia – Pakistan – Mongolia; (E)
G. thomsonii	Thomson's gazelle	Sudan – Tanzania

Addra gazelle
(Gazella dama)

Procapra; Chinese gazelles; Mongolia – Tibet; steppe.

P. gutturosa	Mongolian gazelle	Mongolia, etc.
P. picticaudata	Tibetan gazelle	Tibetan Plateau
P. przewalskii	Przewalski's gazelle	C China

Pantholops

P. hodgsonii	Chiru (Tibetan antelope)	Tibetan Plateau

Saiga

S. tatarica	Saiga	S Russia – Mongolia; steppe

Nemorhaedus; gorals; E Asia; montane forest.

N. cranbrooki	Red goral	N Burma; V
N. goral	Common goral	Himalayas – SE Siberia

Capricornis; serows; E, SE Asia; forest.

C. crispus	Japanese serow	Japan, Taiwan; V
C. sumatraensis	Mainland serow	Kashmir – C China – Sumatra; (E)

Oreamnos

O. americanus	Mountain goat	NW USA – S Alaska

Rupicapra; chamois; Europe, montane; ref. 8.12; (E/V/R).

R. pyrenaica	Pyrenean chamois	Iberia, Pyrenees, C Italy
R. rupicapra	Alpine chamois	Alps – Caucasus, [New Zealand]

Ovibos

O. moschatus	Musk ox	Alaska – Greenland; [Norway, Spitzbergen, N Siberia]; tundra

Budorcas

B. taxicolor	Takin	E Himalayas, SW China; montane forest; (R/I)

Hemitragus; tahrs; SW, S Asia; montane.

H. hylocrius	Nilgiri tahr	S India; V
H. jayakari	Arabian tahr	Oman; E
H. jemlahicus	Himalayan tahr	Himalayas, [New Zealand, California, S Africa]

Capra; goats; S Europe – C Asia – NE Africa; montane.

C. aegagrus (*hircus*)	Wild goat (Bezoar)	Pakistan – Asia Minor, Crete; ancestor of domestic goat *C. hircus*
C. caucasica	West Caucasian tur	W Caucasus
C. cylindricornis	East Caucasian tur	E Caucasus
C. falconeri	Markhor	W Himalayas, Afghanistan, etc.; E/V
C. ibex (*sibirica*) (*walie*)	Ibex	Alps – C Asia – Ethiopia; (E)
C. pyrenaica	Spanish ibex	Spain; (E)

Ammotragus; (*Capra*).

A. lervia	Barbary sheep (Aoudad)	Sahara, etc., [S, SW USA]; (V)

Pseudois

P. nayaur	Bharal (Blue sheep)	Himalayas, W China; montane
P. schaeferi	Dwarf blue sheep	Upper Yangtze, China

Ovis; sheep; W, C, NE Asia, W N America; montane; classification unstable.

O. ammon	Argali	Altai – Himalayas; V
O. canadensis	American bighorn (Mountain sheep)	SW Canada – W USA – N Mexico
O. dalli	Dall sheep (White sheep)	Alaska – N Br. Colombia
O. nivicola	Siberian bighorn	NE Siberia

Ibex
(Capra ibex)

O. orientalis (*laristanica*) (*musimon*)	Mouflon	Iran – Asia Minor, [Sardinia, Corsica, Cyprus, C Europe]; ancestor of domestic sheep, *O. aries*; (V)
O. vignei	Urial	Kashmir – Iran, Turkestan

ORDER PHOLIDOTA

One family.

Family Manidae

Pangolins (scaly anteaters); 7 species; Africa, S Asia; forest, savanna; arboreal, terrestrial; feed on ants and termites; ref. 8.6.

Manis; (*Paramanis*); Asian pangolins.

M. crassicaudata	Indian pangolin	India, Sri Lanka
M. javanica	Malayan pangolin	Burma – Java, Borneo, Palawan
M. pentadactyla	Chinese pangolin	Nepal – S China, Taiwan

Tree pangolin
(Phataginus tricuspis)

Phataginus; (*Smutsia*, *Uromanis*); African pangolins.

P. gigantea	Giant ground pangolin	Senegal – Uganda – Angola; savanna, forest
P. temminckii	Temminck's ground pangolin	S Sudan – S Africa; savanna
P. tetradactyla (*longicaudata*)	Long-tailed pangolin	Senegal – Uganda – Angola; forest, arboreal
P. tricuspis	Tree pangolin	Senegal – W Kenya – Angola; forest, arboreal

ORDER RODENTIA

Rodents; *c.* 1793 species; worldwide; terrestrial, arboreal, subterranean, aquatic; mainly seed-eaters, also herbivores and insectivores.

Family Aplodontidae

One species.

Aplodontia

A. rufa	Mountain beaver (Sewellel)	NW USA; wet forest; I

Mountain beaver
(Aprodontia rufa)

Family Sciuridae

Squirrels; *c.* 254 species; Eurasia, Africa, N, C, S America; forest, savanna, (steppe).

*Tassel-eared squirrel
(Sciurus aberti)*

Subfamily Sciurinae

Tree and ground squirrels; *c.* 216 species.

Sciurus; (*Guerlinguetus, Neosciurus*); Palaearctic and American tree squirrels; forest; arboreal.

S. aberti (*kaibabensis*)	Tassel-eared squirrel	Colorado – NW Mexico, Arizona; montane pine forest
S. aestuans (*gilvigularis*)		N Argentina – Venezuela
S. alleni	Allen's squirrel	NE Mexico
S. anomalus	Persian squirrel	Iran, Asia Minor, etc.
S. arizonensis	Arizona grey squirrel	Arizona, N Mexico; montane
S. aureogaster (*griseoflavus*) (*nelsoni*) (*poliopus*)	Mexican grey squirrel	C Mexico – Guatemala
S. carolinensis	Eastern grey squirrel	E, C USA, SE Canada, [Britain, Ireland, S Africa]; dec. forest
S. colliaei	Collie's squirrel	NW Mexico; montane
S. deppei	Deppe's squirrel	E Mexico – Costa Rica
S. flammifer		Venezuela
S. granatensis	Tropical red squirrel	Ecuador, Venezuela – Costa Rica
S. griseus	Western grey squirrel	Washington – California
S. ignitus		N Argentina – Peru, W Brazil
S. igniventris		Colombia – Peru, N Brazil
S. lis	Japanese squirrel	Japan, except Hokkaido
S. nayaritensis (*apache*)	Nayarit squirrel	NW Mexico – S Arizona
S. niger	Eastern fox squirrel	E, C USA; open forest; (E)
S. oculatus	Peters' squirrel	C Mexico
S. pucheranii		Colombia
S. pyrrhinus		E Peru
S. richmondi	Richmond's squirrel	Nicaragua
S. sanborni		SE Peru
S. spadiceus (*langsdorffi*) (*pyrrhonotus*)		Amazon Basin
S. stramineus		NE Peru, SE Ecuador
S. variegatoides (*goldmani*)	Variegated squirrel	SE Mexico – Panama
S. vulgaris	Eurasian red squirrel	Ireland – Hokkaido; con. forest
S. yucatanensis	Yucatan squirrel	Yucatan, Guatemala

Syntheosciurus; Central America; montane forest.

S. brochus (*poasensis*)	Panama mountain squirrel	Panama, Costa Rica; K

Microsciurus; American pygmy squirrels.

M. alfari	Alfaro's pygmy squirrel	S Nicaragua – Panama
M. flaviventer		N Brazil, Peru
M. mimulus (*isthmius*) (*boquetensis*)	Cloud-forest pygmy squirrel	Panama – Ecuador
M. santanderensis		Colombia

Sciurillus

S. pusillus	South American pygmy squirrel	S America N of Amazon; forest

Prosciurillus; Sulawesi dwarf squirrels; Sulawesi.

P. abstrusus	SE Sulawesi
P. leucomus	Sulawesi
P. murinus	Sulawesi

Rheithrosciurus

R. macrotis	Tufted ground squirrel	Borneo; forest

Tamiasciurus; American red squirrels (chickarees); N America; coniferous forest.

T. douglasii	Douglas' squirrel	W coast USA
T. hudsonicus (*fremonti*)	American red squirrel	Canada, NE USA, Rockies
T. mearnsi		Baja California

Funambulus; palm squirrels (Indian striped squirrels); India etc.; forest, scrub; terrestrial.

F. layardi	Layard's striped squirrel	Sri Lanka, S India
F. palmarum	Indian palm squirrel	Sri Lanka, India
F. pennantii	Northern palm squirrel	N, C India, Pakistan, Nepal
F. sublineatus	Dusky striped squirrel	Sri Lanka, S India
F. tristriatus	Jungle striped squirrel	SW India

Ratufa; oriental giant squirrels; SE Asia; forest.

R. affinis	Cream-coloured giant squirrel	Malaya, Sumatra, Borneo
R. bicolor	Black giant squirrel	Nepal – S China – Java
R. indica	Indian giant squirrel	S, C India
R. macroura	Grizzled Indian squirrel	S India, Sri Lanka

Giant forest squirrel (Protoxerus stangeri)

Protoxerus; African giant squirrels; W, C, E Africa; forest.

P. aubinni	Slender-tailed giant squirrel	Liberia – Ghana
P. stangeri	Giant forest squirrel	Sierra Leone – Kenya – Angola

Epixerus; African palm squirrels; high forest.

E. ebii	Temminck's giant squirrel	Sierra Leone – Ghana
E. wilsoni		Cameroun – R Zaire

Funisciurus; African striped squirrels, rope squirrels; Africa; forest.

F. anerythrus	Thomas' tree squirrel	Senegal, Nigeria – Uganda, Zaire
F. bayonii	Bocage's tree squirrel	NE Angola, SW Zaire
F. carruthersi	Mountain tree squirrel	Ruwenzori – Burundi; montane
F. congicus	Kuhl's tree squirrel	R Zaire – SW Africa
F. isabella	Gray's four-striped squirrel	Cameroun – R Zaire
F. lemniscatus	Leconte's four-striped squirrel	R Sanaga – R Zaire
F. leucogenys	Orange-headed squirrel	Ghana – Cent. Af. Rep. – Rio Muni
F. pyrrhopus	Cuvier's tree squirrel	Gambia – Uganda – Angola
F. substriatus	De Winton's tree squirrel	Ivory Coast – Nigeria

Paraxerus; (*Aethosciurus*, *Montisciurus*); African bush squirrels; Africa; savanna, dry forest.

P. alexandri	Alexander's bush squirrel	NE Zaire, Uganda
P. boehmi	Boehm's bush squirrel	E, C Africa
P. cepapi	Smith's bush squirrel (S African tree squirrel)	S Angola – S Tanzania – Transvaal
P. cooperi	Cooper's green squirrel	Cameroun
P. flavivittis	Striped bush squirrel	N Mozambique – S Kenya
P. lucifer	Black and red bush squirrel	N Malawi, SW Tanzania; montane
P. ochraceus	Huet's bush squirrel	Tanzania – S Sudan
P. palliatus	Red bush squirrel	Natal – Somalia
P. poensis	Small green squirrel	Sierra Leone – R Zaire
P. vexillarius	Swynnerton's bush squirrel	C, E Tanzania; montane
P. vincenti	Vincent's bush squirrel	N Mozambique; (in *P. vexillarius*?)

Heliosciurus; sun squirrels; Africa; forest; arboreal.

H. gambianus	Gambian sun squirrel	Senegal – Ethiopia – Zambia; savanna, secondary forest
H. mutabilis	Southern sun squirrel	Zimbabwe – S Tanzania
H. rufobrachium	Red-legged sun squirrel	W Africa
H. ruwenzorii	Ruwenzori sun squirrel	E Zaire, etc.; montane forest
H. undulatus	Eastern sun squirrel	E Africa

Hyosciurus; Sulawesi long-nosed squirrels.

H. heinrichi		Sulawesi; montane forest
H. ileile		Sulawesi; lowland forest

Myosciurus

M. pumilio	African pygmy squirrel	SE Nigeria – Gabon; forest

Callosciurus; oriental tree squirrels; SE Asia.

C. adamsi	Ear-spot squirrel	N, NW Borneo
C. albiculus		N Sumatra
C. baluensis	Kinabalu squirrel	C, N Borneo; montane
C. caniceps	Grey-bellied squirrel (Golden-backed squirrel)	Thailand – Malaya
C. erythraeus (*flavimanus*) (*quinquestriatus*)	Belly-banded squirrel	Sikkim – S China – Malaya, Taiwan
C. finlaysonii (*ferrugineus*)	Finlayson's squirrel (Variable squirrel)	E Burma – Indochina
C. inornatus		Indochina
C. melanogaster		Mentawai Is, Sumatra
C. nigrovittatus	Black-banded squirrel	Malaya, Sumatra, Java
C. notatus	Plantain squirrel	Malaya – Java, Borneo; secondary forest
C. orestes (*canalvus*)	Bornean black-banded squirrel	Borneo; montane
C. phayrei		Burma
C. prevostii	Prevost's squirrel	Malaya, Sumatra, Borneo, (Sulawesi)
C. pygerythrus	Irrawaddy squirrel	Nepal – Burma
C. simus	Sculptor squirrel	Mt Kinabalu, Borneo

Tamiops; (*Callosciurus*); Oriental striped squirrels; S China, etc.

T. mcclellandii	Himalayan striped squirrel	Nepal – S China – Malaya
T. maritimus		S China, Indochina, Taiwan
T. rodolphei	Cambodian striped squirrel	Indochina
T. swinhoei	Swinhoe's striped squirrel	NE China – N Vietnam

Rubrisciurus; (*Callosciurus*).

R. rubriventer	Sulawesi giant squirrel	Sulawesi

Sundasciurus; (*Callosciurus*); Sunda squirrels; SE Asia; forest.

S. brookei	Brooke's squirrel	C, N Borneo
S. hippurus	Horse-tailed squirrel	Malaya, Sumatra, Borneo
S. jentinki	Jentink's squirrel	Borneo
S. juvencus		Palawan, Philippines
S. lowii	Low's squirrel	Malaya, Sumatra, Borneo
S. mindanensis		Mindanao, Philippines
S. moellendorffi (*albicauda*) (*hoogstraali*)		Calamian Is, Philippines
S. philippinensis		Basilan, Mindanao etc., Philippines

S. rabori		Palawan
S. samarensis		Samar etc., Philippines
S. steerii		Palawan, Philippines
S. tenuis	Slender squirrel	Malaya, Sumatra, Borneo

Menetes

M. berdmorei	Berdmore's squirrel (Indochinese ground squirrel)	Burma – Indochina

Rhinosciurus

R. laticaudatus	Shrew-faced squirrel (Long-nosed squirrel)	Malaya, Sumatra, Borneo; forest

Lariscus

L. hosei	Four-striped ground squirrel	N, NW Borneo; forest
L. insignis (*niobe*)	Three-striped ground squirrel	Malaya – Java, Borneo; forest
L. obscurus	Mentawai ground squirrel	Mentawai Is, Sumatra

Three-stiped ground squirrel
(Lariscus insignis)

Dremomys

D. everetti	Bornean mountain ground squirrel	N, NW Borneo; montane forest
D. gularis	Red-throated squirrel	Vietnam
D. lokriah	Orange-bellied Himalayan squirrel	Nepal – N Burma; montane forest
D. pernyi (*pyrrhomeris*)	Perny's long-nosed squirrel	S China – Burma, Taiwan
D. rufigenis	Red-cheeked squirrel	Indochina – Malaya

Sciurotamias

S. davidianus	Père David's rock squirrel	Hebei – Sichuan, China
S. forresti	Forrest's rock squirrel	Yunnan, China

Nannosciurus

N. melanotis	Black-eared pygmy squirrel	Sumatra, Java, Borneo

Exilisciurus; ref. 9.3.

E. concinnus (*luncefordi*) (*samaricus*) (*surrutilus*)	Philippine pygmy squirrel	Mindanao etc., Philippines
E. exilis	Plain pygmy squirrel	Borneo; lowland forest
E. whiteheadi	Whitehead's pygmy squirrel	NW Borneo; montane forest

Atlantoxerus

A. getulus	Barbary ground squirrel	Morocco, Algeria

Xerus; (*Euxerus*, *Geosciurus*); African ground squirrels; Africa S of Sahara; savanna, steppe; terrestrial.

X. erythropus	Geoffroy's ground squirrel	Morocco, Senegal – Kenya, Ethiopia
X. inauris	Cape ground squirrel	Africa S of Zambezi
X. princeps	Kaokoveld ground squirrel (Mountain ground squirrel)	S W Africa, S Angola
X. rutilus	Unstriped ground squirrel	Ethiopia – N Tanzania

Spermophilopsis

S. leptodactylus	Long-clawed ground squirrel	Russian Turkestan etc.; desert

Marmota; marmots; Palaearctic, N America; alpine talus, grassland, (forest).

M. baibacina	Steppe marmot	Kazakhstan, etc.
M. bobak	Bobak marmot	S Russia
M. broweri	Alaska marmot	N Alaska; vicariant of *M. camtschatica*
M. caligata	Hoary marmot	Alaska – Idaho; montane
M. camtschatica	Black-capped marmot	NE Siberia; montane
M. caudata	Long-tailed marmot	Tien Shan – Kashmir; montane
M. flaviventris	Yellow-bellied marmot	W USA; montane
M. himalayana	Himalayan marmot	Himalayas
M. marmota	Alpine marmot	C Europe; montane
M. menzbieri	Menzbier's marmot	W Tien Shan; V
M. monax	Woodchuck	Alaska – Labrador, E USA; forest
M. olympus	Olympic marmot	Olympic Peninsula, Washington, USA
M. sibirica	Siberian marmot	SE Siberia, N Mongolia, etc.
M. vancouverensis	Vancouver marmot	Vancouver I, Canada; E

Cynomys; prairie dogs; C, W USA – C Mexico; grassland.

C. gunnisoni	Gunnison's prairie dog	Colorado – Arizona; high grassland
C. leucurus	White-tailed prairie dog	Wyoming, etc.
C. ludovicianus	Black-tailed prairie dog	C USA; grassland
C. mexicanus	Mexican prairie dog	C Mexico
C. parvidens	Utah prairie dog	Utah; V

Black-tailed prairie dog (Cynomys ludovicianus)

Spermophilus; (*Callospermophilus*, *Citellus*, *Otospermophilus*); ground squirrels, sousliks; N America, N Eurasia; grassland, steppe, (open forest).

S. adocetus	Tropical ground squirrel	C Mexico
S. alaschanicus	Alashan ground squirrel	W Gansu, S Mongolia
S. annulatus	Ring-tailed ground squirrel	W Mexico
S. armatus	Uinta ground squirrel	Rocky Mts (USA)

S. atricapillus	Baja California rock squirrel	Baja California
S. beecheyi	California ground squirrel	California, W Oregon
S. beldingi	Belding's ground squirrel	W USA
S. brunneus	Idaho ground squirrel	W Idaho
S. citellus	European souslik	SE Europe
S. columbianus	Columbian ground squirrel	NW USA, SW Canada
S. dauricus	Daurian ground squirrel	E Mongolia, Transbaikalia, N China
S. elegans (*richardsonii*)		C Rockies of USA
S. erythrogenys (*major*)		E Kazakhstan – Mongolia
S. franklinii	Franklin's ground squirrel	C USA, SC Canada; tall grass
S. fulvus	Large-toothed souslik	Turkestan
S. lateralis	Golden-mantled ground squirrel	W USA, SW Canada; montane forest, scrub
S. madrensis	Sierra Madre ground squirrel	Chihuahua, Mexico
S. major	Russet souslik	S Russia – SW Siberia
S. mexicanus	Mexican ground squirrel	E Mexico, Texas
S. mohavensis	Mohave ground squirrel	Mohave Desert, S California
S. musicus (*citellus*)		N Caucasus
S. parryii	Arctic souslik (Arctic ground squirrel)	NE Siberia; Alaska – Hudson Bay; tundra
S. perotensis	Perote ground squirrel	Veracruz, Mexico
S. pygmaeus	Little souslik	S Russia, Kazakhstan
S. relictus	Tien Shan souslik	Tien Shan Mts
S. richardsoni	Richardson's ground squirrel	NW USA, SW Canada; grassland
S. saturatus	Cascade golden-mantled ground squirrel	Cascade Mts, Washington, Br. Columbia
S. spilosoma	Spotted ground squirrel	SC USA, N Mexico
S. suslicus	Spotted souslik	E Europe, S Russia
S. tereticaudus	Round-tailed ground squirrel	SW USA, NW Mexico; desert, scrub
S. townsendii	Townsend's ground squirrel	W USA; steppe
S. tridecemlineatus	Thirteen-lined ground squirrel	C USA, SC Canada; short grass
S. undulatus	Long-tailed souslik	Tien Shan – R Lena
S. variegatus	Rock squirrel	SW USA, Mexico; dry steppe, desert
S. washingtoni	Washington ground squirrel	Washington, Oregon
S.xanthoprymnus (*citellus*)		Asia Minor, etc.

Ammospermophilus; antelope-squirrels; SW USA, N Mexico; desert, steppe.

A. harrisii	Harris' antelope-squirrel	Arizona, N Mexico
A. interpres	Texas antelope-squirrel	New Mexico – N Mexico
A. leucurus (*insularis*)	White-tailed antelope-squirrel	SW USA
A. nelsoni	Nelson's antelope-squirrel	California

Tamias; (*Eutamias*); chipmunks; N America, N Eurasia.

T. alpinus	Alpine chipmunk	California; high montane
T. amoenus	Yellow-pine chipmunk	NW USA, SW Canada
T. bulleri	Buller's chipmunk	N Mexico
T. canipes	Grey-footed chipmunk	New Mexico, Texas
T. cinereicollis	Grey-collared chipmunk	Arizona, New Mexico; montane forest
T. dorsalis	Cliff chipmunk	SW USA, N Mexico; pine, juniper forest
T. merriami	Merriam's chipmunk	S California; scrub, forest
T. minimus	Least chipmunk	W USA, S, C Canada; forest, scrub
T. obscurus (*merriami*)	Chaparral chipmunk (California chipmunk)	S California
T. palmeri	Palmer's chipmunk	Nevada; montane forest
T. panamintinus	Panamint chipmunk	Arizona, California; scrub
T. quadrimaculatus	Long-eared chipmunk	California
T. quadrivittatus	Colorado chipmunk	SW USA; montane forest
T. ruficaudus	Red-tailed chipmunk	NW USA, SW Canada; montane forest
T. rufus		W Colorado, etc.; ref. 9.5
T. sibiricus	Siberian chipmunk	Siberia, W China, Hokkaido
T. sonomae	Sonoma chipmunk	NW California; scrub
T. speciosus	Lodgepole chipmunk	California
T. striatus	Eastern chipmunk	E USA, SE Canada; forest
T. townsendii (*ochrogenys*) (*senex*) (*siskiyou*)	Townsend's chipmunk	W coast USA; ref. 9.6
T. umbrinus	Uinta chipmunk	SW USA; montane forest

Eastern chipmunk
(Tamius striatus)

Subfamily Pteromyinae

Flying squirrels (gliding squirrels); *c.* 38 species; SE Asia, (N Eurasia, N America).

Petaurista; giant flying squirrels; E, SE Asia; forest.

P. alborufus	Red and white flying squirrel	S China, Taiwan
P. elegans	Spotted giant flying squirrel	Nepal – Java, Borneo
P. leucogenys	Japanese giant flying squirrel	Japan except Hokkaido

Southern flying squirrel
(Glaucomyx volans)

P. magnificus	Hodgson's flying squirrel	Nepal, Sikkim, etc.
P. nobilis	Bhutan flying squirrel	Nepal – Bhutan
P. petaurista	Red giant flying squirrel	Kashmir – Malaya – Java, Borneo
P. philippensis (*petaurista*)	Indian giant flying squirrel	Sri Lanka, India – S China, Taiwan
P. xanthotis (*leucogenys*)	Chinese giant flying squirrel	Gansu – Yunnan

Biswamoyopterus

B. biswasi	Namdapha flying squirrel	Arunachal Pradesh, NE India; ref. 9.7

Eupetaurus

E. cinereus	Woolly flying squirrel	Kashmir; montane coniferous forest, rocks

Pteromys

P. momonga	Small Japanese flying squirrel	Honshu, Kyushu, Japan
P. volans	Siberian flying squirrel	Finland – Korea, Hokkaido; coniferous forest

Glaucomys; American flying squirrels; N America; forest.

G. sabrinus	Northern flying squirrel	Canada, W USA
G. volans	Southern flying squirrel	E USA – Honduras

Aeromys

A. tephromelas (*phaeomelas*)	Black flying squirrel	Malaya, Sumatra, Borneo
A. thomasi	Thomas' flying squirrel	Borneo

Hylopetes; pygmy flying squirrels; SE Asia; forest.

H. alboniger	Particoloured flying squirrel	Nepal – Indochina
H. baberi (*fimbriatus*)		NE Afghanistan – Kashmir; montane; ref. 9.8
H. fimbriatus	Kashmir pygmy flying squirrel	Kashmir etc.
H. electilis	Hainan flying squirrel	Hainan, China; formerly in *Petinomys*
H. lepidus	Grey-cheeked flying squirrel	Malaya – Java, Borneo
H. mindanensis		Mindanao, Philippines
H. nigripes		Palawan, Philippines
H. phayrei	Phayre's flying squirrel	Burma, Thailand, Laos
H. platyurus	Grey-cheeked flying squirrel	Malaya – Java, Borneo
H. sipora		Sipora I, Mentawai Is

H. spadiceus	Red-cheeked flying squirrel	Burma – Malaya, Sumatra, Borneo

Petinomys

P. bartelsi		Java
P. crinitus		Mindanao etc., Philippines
P. fuscocapillus	Travancore flying squirrel	S India, Sri Lanka
P. genibarbis	Whiskered flying squirrel	Malaya – Borneo
P. hageni		Borneo, Sumatra
P. lugens (*hageni*)		Mentawai Is
P. setosus (*morrisi*)	White-bellied flying squirrel	Burma, Malaya, Sumatra, Borneo
P. vordermanni	Vordermann's flying squirrel	Malaya, Borneo

Aeretes

A. melanopterus	North Chinese flying squirrel	NE China, Sichuan

Trogopterus

T. xanthipes	Complex-toothed flying squirrel	China

Belomys

B. pearsonii	Hairy-footed flying squirrel	E Nepal – Indochina, Taiwan

Pteromyscus

P. pulverulentus	Smoky flying squirrel	S Thailand – Sumatra, Borneo

Petaurillus; pygmy flying squirrels, SE Asia.

P. emiliae		Sarawak, Borneo
P. hosei (*kinlochii*)		Sarawak, Borneo, Malaya

Iomys

I. horsfieldii	Horsfield's flying squirrel	Malaya – Java, Borneo
I. sipora		Mentawai Is
I. winstoni		N Sumatra

Family Geomyidae

Pocket gophers; *c.* 35 species; USA (especially SW) to Colombia; forest – steppe; subterranean.

Geomys

G. arenarius	Desert pocket gopher	S New Mexico, etc.
G. breviceps (*bursarius*)	Baird's pocket gopher	Louisiana

Botta's pocket gopher
(*Thomomys bottae*)

G. bursarius	Plains pocket gopher	C USA
G. personatus	Texas pocket gopher	S Texas, NE Mexico
G. pinetis	Southeastern pocket gopher	Alabama – C Florida
(*colonus*)		
(*cumberlandius*)		
(*fontanelus*)		
G. tropicalis	Tropical pocket gopher	S Tamaulipas, E Mexico
(*personatus*)		

Thomomys

T. bottae	Botta's pocket gopher	S USA – N Mexico
T. bulbivorus	Camas pocket gopher	NW Oregon
T. clusius		Wyoming
T. idahoensis	Idaho pocket gopher	Idaho, etc.
(*talpoides*)		
T. mazama	Western pocket gopher	Oregon, N California
T. monticola	Mountain pocket gopher	California
T. talpoides	Northern pocket gopher	NW USA, SW Canada
T. umbrinus	Southern pocket gopher	SW USA, Mexico
(*townsendi*)		

Pappogeomys; (*Cratogeomys*); Mexico, (S USA).

P. alcorni	Alcorn's pocket gopher	Jalisco, Mexico
P. bulleri	Buller's pocket gopher	SW Mexico
P. castanops	Yellow-faced pocket gopher	SE Colorado – N Mexico
P. fumosus	Smoky pocket gopher	Colima, SW Mexico
P. gymnurus	Llano pocket gopher	C Mexico
P. merriami	Merriam's pocket gopher	C Mexico
P. neglectus	Queretaro pocket gopher	Queretaro, C Mexico
P. tylorhinus	Taylor's pocket gopher	C Mexico
P. zinseri	Zinser's pocket gopher	Jalisco, C Mexico

Orthogeomys; (*Heterogeomys, Macrogeomys*); Mexico – Colombia.

O. cavator	Chiriqui pocket gopher	W Panama, Costa Rica
O. cherriei	Cherrie's pocket gopher	Costa Rica
O. cuniculus	Oaxacan pocket gopher	Oaxaca, Mexico
O. dariensis	Darien pocket gopher	E Panama – Colombia
O. grandis	Large pocket gopher	Honduras – Guerrero, Mexico
O. heterodus	Variable pocket gopher	Costa Rica
O. hispidus	Hispid pocket gopher	E Mexico – Guatemala
O. lanius	Big pocket gopher	Veracruz, Mexico
O. matagalpae	Nicaraguan pocket gopher	Nicaragua, Honduras
O. pygacanthis	El Salvador pocket gopher	El Salvador
O. underwoodi	Underwood's pocket gopher	Costa Rica

Zygogeomys

Z. trichopus	Michoacan pocket gopher	Michoacan, SW Mexico

Family Heteromyidae

Pocket mice, kangaroo-rats, etc.; *c.* 60 species; N, (C, S) America; desert, steppe, (forest).

*Desert pocket mouse
(Chaetodipus pencillatus)*

Perognathus; pocket mice; Mexico, SW, (NW) USA; desert, steppe.

P. alticola	White-eared pocket mouse	S California; V
P. amplus	Arizona pocket mouse	Arizona, etc.
P. anthonyi	Anthony's pocket mouse	Cerros I, Baja California
P. arenarius	Little desert pocket mouse	Baja California
P. artus	Narrow-skulled pocket mouse	NW Mexico
P. dalquesti	Dalquest's pocket mouse	Baja California
P. fasciatus	Olive-backed pocket mouse	Colorado – Saskatchewan
P. flavescens (*apache*)	Plains pocket mouse	Arizona – N Texas – Minnesota
P. flavus (*merriami*)	Silky pocket mouse	W Nebraska – C Mexico
P. goldmani	Goldman's pocket mouse	NW Mexico
P. inornatus	San Joaquin pocket mouse	California; E
P. lineatus	Lined pocket mouse	C Mexico
P. longimembris	Little pocket mouse	Utah – Baja California; E
P. parvus	Great Basin pocket mouse	Nevada – S Br. Colombia
P. pernix	Sinaloan pocket mouse	NW Mexico
P. xanthonotus	Yellow-eared pocket mouse	C California

Chaetodipus; (*Perognathus*); spiny-rumped pocket mice.

C. baileyi	Bailey's pocket mouse	Baja California, Arizona, Sonora
C. californicus	California pocket mouse	California
C. fallax	San Diego pocket mouse	S California, N Baja California
C. formosus	Long-tailed pocket mouse	Utah – Baja California
C. hispidus	Hispid pocket mouse	N Dakota – C Mexico
C. intermedius	Rock pocket mouse	Arizona – SW Texas, N Mexico
C. nelsoni	Nelson's pocket mouse	N Mexico, SW Texas
C. penicillatus	Desert pocket mouse	S California – C Mexico
C. spinatus	Spiny pocket mouse	Baja California, S California

Microdipodops; kangaroo-mice; Nevada, etc; sand-steppes.

M. megacephalus	Dark kangaroo-mouse	Nevada, etc.
M. pallidus	Pale kangaroo-mouse	SW Nevada

Dipodomys; kangaroo-rats; SW USA – Mexico; desert, steppe.

D. agilis (*paralius*) (*peninsularis*)	Agile kangaroo-rat (Pacific kangaroo-rat)	C California – Baja California
D. californicus	Californian kangaroo-rat	N California, S Oregon

D. compactus	Gulf coast kangaroo-rat	S Texas, NE Mexico
D. deserti	Desert kangaroo-rat	Nevada – NW Mexico
D. elator	Texas kangaroo-rat	N Texas; R
D. elephantinus	Big-eared kangaroo-rat	WC California
D. gravipes	San Quintin kangaroo-rat	NW Baja California; E
D. heermanni	Heermann's kangaroo-rat	California; (E)
D. ingens	Giant kangaroo-rat	S California; I
D. insularis	San Jose kangaroo-rat	San Jose I, Baja California
D. margaritae (*merriami*)	Santa Margarita kangaroo-rat	Santa Margarita I, Baja California
D. merriami	Merriam's kangaroo-rat	Nevada – C Mexico
D. microps	Chisel-toothed kangaroo-rat	Nevada, etc.
D. nelsoni	Nelson's kangaroo-rat	NC Mexico; close to *D. spectabilis*
D. nitratoides	Fresno kangaroo-rat	C California; E
D. ordii	Ord's kangaroo-rat	W USA, NE Mexico
D. panamintinus	Panamint kangaroo-rat	California, SW Nevada; montane
D. phillipsi (*ornatus*)	Phillips' kangaroo-rat	C Mexico
D. spectabilis	Banner-tailed kangaroo-rat	New Mexico – C Mexico
D. stephensi	Stephens' kangaroo-rat	S California; I
D. venustus	Narrow-faced kangaroo-rat	WC California

Liomys; spiny pocket-mice; Mexico – Panama; arid scrub, steppe.

L. adspersus	Panamanian spiny pocket mouse	C Panama
L. irroratus	Mexican spiny pocket mouse	S Texas – S Mexico
L. pictus (*annectens*)	Painted spiny pocket mouse	W, S Mexico
L. salvini (*crispus*)	Salvin's spiny pocket mouse	S Mexico – Costa Rica
L. spectabilis	Jaliscan spiny pocket mouse	Jalisco, WC Mexico

Heteromys; forest spiny pocket mice; S Mexico – Ecuador; forest.

H. anomalus	Trinidad spiny pocket mouse	Colombia, Venezuela, Trinidad
H. australis	Southern spiny pocket mouse	E Panama – Ecuador
H. desmarestianus (*lepturus*) (*longicaudatus*) (*nigricaudatus*) (*temporalis*)	Desmarest's spiny pocket mouse	S Mexico – Colombia; ref. 9.9
H. gaumeri	Gaumer's spiny pocket mouse	Yucatan, Mexico

H. goldmani	Goldman's spiny pocket mouse	Guatemala, Chiapas
H. nelsoni	Nelson's spiny pocket mouse	Chiapas, S Mexico
H. oresterus	Mountain spiny pocket mouse	Costa Rica

Family Castoridae

Castor; beavers; 2 species; freshwater in deciduous forest.

| *C. canadensis* | American beaver | N America, [Finland, Tierra del Fuego] |
| *C. fiber* (*albicus*) | Eurasian beaver | Europe – Altai |

*American beaver
(Castor canadensis)*

Family Anomaluridae

Scaly-tailed squirrels; 7 species; W, C Africa; forest; arboreal.

Anomalurus; (*Anomalurops*); large scaly-tailed squirrels; W, C Africa.

A. beecrofti	Beecroft's flying squirrel	W Africa – Zaire
A. derbianus	Lord Derby's flying squirrel	Sierra Leone – Mozambique, Angola
A. pelii	Pel's flying squirrel	Sierra Leone – Ghana
A. pusillus	Little flying squirrel	Zaire, S Cameroun, Gabon

Idiurus; pygmy scaly-tailed squirrels; W, C Africa.

| *I. macrotis* | Long-eared flying squirrel | W Africa – Zaire |
| *I. zenkeri* | Zenker's flying squirrel | Cameroun, Zaire |

*Beecroft's flying squirrel
(Anomalurus beecrofti)*

Zenkerella

| *Z. insignis* | Flightless scaly-tailed squirrel | Cameroun, Rio Muni, ? Gabon; forest |

Family Pedetidae

Pedetes

| *P. capensis* | Spring hare (Springhaas) | S Africa – S Kenya; steppe |

Family Muridae

Mice, rats, voles, gerbils, hamsters, etc.; *c.* 1160 species; worldwide in all terrestrial (and freshwater) habitats; feed mainly on seeds, also grass, insects, etc.; classification very provisional – the groups recognized here as subfamilies are often treated as independent families, especially the Cricetidae and Spalacidae.

*Spring hare
(Pedetes capensis)*

Subfamily Hesperomyinae

New-world mice and rats; *c.* 372 species; N, C, S America; desert – forest; classification of many Central and South American representatives very provisional.

*Marsh rice rat
(Oryzomys palustris)*

Oryzomys; (*Melanomys, Nectomys, Nesoryzomys, Oecomys, Oligoryzomys, Sigmodontomys*); rice rats; S USA – Tierra del Fuego; grassland, marshes, scrub, (forest).

O. albigularis
  (*devius*)
  (*pirrensis*)
	Tomes' rice rat	Bolivia – Costa Rica;
		  montane

O. alfari
  (*russulus*)
	Alfaro's water rat	Honduras – Venezuela,
		  Ecuador

O. alfaroi	Alfaro's rice rat	C Mexico – Ecuador
O. altissimus		Peru, Ecuador; montane
O. andinus		N Peru
O. aphrastus	Harris's rice rat	Costa Rica
O. arenalis		NE Peru
O. auriventer		Peru, Ecuador
O. balneator		Ecuador, E of Andes
O. bauri		Sante Fe I, Galapagos; (in *O. galapagoensis*?)

O. bicolor
  (*endersi*)
  (*phaeotis*)
  (*trabeatus*)
		Bolivia, Amazon Basin –
		  Panama; forest

O. bombycinus	Silky rice rat (Long-whiskered rice rat)	Nicaragua – N Ecuador
O. buccinatus		Paraguay, N Argentina
O. caliginosus	Dusky rice rat	Honduras – Ecuador
O. capito		Brazil, N Argentina – Venezuela
O. caudatus		Oaxaca, S Mexico
O. chacoensis		C Brazil – Argentina; ref. 9.81
O. cleberi		C Brazil; close to *O. bicolor*; ref. 9.10

O. concolor
  (*trinitatis*)
		N Argentina, Bolivia –
		  Costa Rica

O. couesi
  (*cozumelae*)
  (*palustris*)
  (*gatunensis*)
  (? *peninsulae*)
		NW Colombia – Texas;
		  Jamaica; ref. 9.83

O. delicatus
  (*microtis*)
		Colombia – Guyana,
		  N Brazil

O. delticola		NE Argentina, Uruguay
O. dimidiatus	Thomas's water rat	Nicaragua
O. flavescens		S Brazil – N Argentina
O. fulgens	Thomas's rice rat	C Mexico
O. fulvescens	Pygmy rice rat	C Mexico – Venezuela

O. galapagoensis †		San Cristobal I,
		  Galapagos; probably
		  extinct

O. gorgasi		NW Colombia; forest
O. hammondi		NW Ecuador

O. intectus		Colombia; montane
O. kelloggi		SE Brazil
O. lamia		SE Brazil
O. longicaudatus		Tierra del Fuego – Peru
O. macconnelli		Surinam – Ecuador
O. melanostoma		E Peru
O. melanotis	Black-eared rice rat	Mexico, El Salvador
O. microtis (*fornesi*) (? *chaparensis*)		Bolivia, N Argentina; ref. 9.19
O. minutus		Peru – Colombia; montane
O. munchiquensis		W Colombia
O. nelsoni	Nelson's rice rat	Maria Madre I, W Mexico
O. nigripes		E Brazil – Argentina
O. nitidus		Ecuador – N Argentina
O. palustris (*argentatus*)	Marsh rice rat	SE USA; marshes; ref. 9.11; (I)
O. polius		N Peru
O. ratticeps		SE Brazil, Paraguay, N Argentina
O. rivularis (*couesi*)		NW Ecuador
O. robustulus		E Ecuador
O. spodiurus		Ecuador
O. subflavus		E Brazil, Guianas
O. talamancae (*capito*)		Costa Rica – Ecuador; ref. 9.85
O. utiaritensis		C Brazil
O. victus †	St Vincent rice rat	St Vincent I, W Indies; extinct
O. xantheolus		Peru
O. yunganus		Bolivia, Peru, Venezuela; ref. 9.120
O. zunigae		Peru; coastal

Nesoryzomys; (*Oryzomys*); Galapagos mice; Galapagos Islands.

N. darwini †	Santa Cruz Is.; probably extinct
N. fernandinae	Fernandina Is.; ref. 9.12
N. indeffesus†	Santa Cruz Is.; probably extinct
N. narboroughi	Fernandina Is.
N. swarthi †	James Is.; probably extinct

Megalomys†; giant rice rats; Lesser Antilles; extinct.

M. desmarestii†	Antillean giant rice rat	Martinique, Lesser Antilles; extinct
M. luciae†	Santa Lucia giant rice rat	Santa Lucia, Lesser Antilles; extinct

Wiedomys; (*Oryzomys*, *Thomasomys*).

W. pyrrhorhinos	Red-nosed mouse	E Brazil; scrub

Neacomys; bristly mice (spiny rice rats); northern S America; forest.

N. guianae	Surinam – S Venezuela
N. spinosus	SW Brazil – Peru – Colombia
N. tenuipes (*pictus*) (*pusillus*)	E Panama – Ecuador, Venezuela

Scolomys

S. melanops	Ecuador spiny mouse	Ecuador

Nectomys; classification provisional – *N. squamipes* is probably composite.

N. parvipes		French Guiana
N. squamipes	South American water rat	S America N of Paraguay & S Brazil

South american water rat (Nectomys squamipes)

Rhipidomys; American climbing mice; S America; forest.

R. latimanus (*fulviventer*) (*venustus*)		Colombia, Venezuela, Ecuador
R. leucodactylus		N Argentina – Venezuela
R. macconnelli		SE Venezuela, etc.
R. maculipes		E Brazil
R. mastacalis (*venezuelae*)		N Brazil, Venezuela, Guianas
R. scandens	Mt Pirri climbing mouse	E Panama
R. sclateri		Venezuela, Guyana, Trinidad

Thomasomys; (*Wilfredomys*); S America; forest.

T. aureus	Peru – Colombia; montane
T. baeops	W Ecuador
T. bombycinus	Colombia; montane
T. cinereiventer	Colombia, Ecuador; montane
T. cinereus	S Ecuador, N Peru
T. daphne	Bolivia, S Peru
T. dorsalis	E Brazil
T. gracilis	Ecuador, Peru
T. hylophilus	NE Colombia, W Venezuela
T. incanus	Peru
T. ischyurus	E Peru, E Ecuador
T. kalinowskii	Peru
T. ladewi	NW Bolivia; montane
T. laniger	W Venezuela, Colombia

T. monochromus (*laniger*)		NE Colombia
T. notatus		SE Peru; montane
T. oenax		S, E Brazil, Uruguay
T. oreas		Bolivia; montane
T. paramorum		Ecuador; montane
T. pictipes		NE Argentina; S Brazil
T. pyrrhonotus		S Ecuador, N Peru
T. rhoadsi		Ecuador
T. rosalinda		N Peru
T. taczanowskii		N Peru
T. vestitus		W Venezuela

Aepeomys; (*Thomasomys*); N Andes.

A. fuscatus (*lugens*)		Colombia
A. lugens		Venezuela – Ecuador

Phaenomys

P. ferrugineus	Rio rice rat	E Brazil

Chilomys

C. instans	Colombian forest mouse	Ecuador – Venezuela; montane forest

Tylomys; American climbing rats; S Mexico – Ecuador; forest.

T. bullaris	Chiapan climbing rat	Chiapas, S Mexico
T. fulviventer (*mirae*)	Fulvous-bellied climbing rat	Panama
T. mirae		Colombia, N Ecuador
T. nudicaudus (*gymnurus*)	Naked-tailed climbing rat	S Mexico – Nicaragua
T. panamensis	Panama climbing rat	E Panama
T. tumbalensis	Tumbala climbing rat	Chiapas, S Mexico
T. watsoni	Watson's climbing rat	Costa Rica, Panama

Ototylomys

O. phyllotis	Big-eared climbing rat	S Mexico – Costa Rica

Nyctomys

N. sumichrasti	Sumichrast's vesper rat	S Mexico – Panama; forest

Otonyctomys

O. hatti	Yucatan vesper rat	Yucatan, Belize, N Guatemala; forest

Rhagomys

R. rufescens		E Brazil; forest

Reithrodontomys; American harvest mice; N, C, (S) America; grassland, forest.

R. brevirostris	Short-nosed harvest mouse	Costa Rica, Nicaragua

R. burti	Sonoran harvest mouse	NW Mexico
R. chrysopsis	Volcano harvest mouse	C Mexico
R. creper	Chiriqui harvest mouse	Costa Rica, W Panama; forest, terrestrial
R. darienensis	Darien harvest mouse	E Panama; forest, arboreal
R. fulvescens	Fulvous harvest mouse	S USA – Nicaragua
R. gracilis	Slender harvest mouse	S Mexico – Costa Rica
R. hirsutus	Hairy harvest mouse	WC Mexico
R. humulis	Eastern harvest mouse	SE USA
R. megalotis	Western harvest mouse	SW Canada – S Mexico; (V)
R. mexicanus	Mexican harvest mouse	C Mexico – Ecuador
R. microdon	Small-toothed harvest mouse	S Mexico – Guatemala
R. montanus	Plains harvest mouse	SC USA, NC Mexico
R. paradoxus		Nicaragua, Costa Rica
R. raviventris	Saltmarsh harvest mouse	San Francisco Bay; E
R. rodriguezi	Rodriguez's harvest mouse	Costa Rica
R. spectabilis	Cozumel harvest mouse	Cozumel I, Mexico
R. sumichrasti	Sumichrast's harvest mouse	C Mexico – Panama
R. tenuirostris	Narrow-nosed harvest mouse	Guatemala; montane

Peromyscus; deer mice, etc.; N, C America; forest – desert.

P. attwateri	Texas mouse	Texas, Oklahoma, etc.; forest, rocks
P. aztecus (*oaxacensis*)	Aztec mouse	C Mexico – Honduras; wet forest
P. boylii	Brush mouse	C, SW USA – Honduras
P. bullatus	Perote mouse	Veracruz, Mexico
P. californicus	California mouse	C, S California; scrub
P. caniceps	Burt's deer mouse	Monserrate I, Baja California
P. crinitus	Canyon mouse	Oregon – Colorado – NW Mexico; desert
P. dickeyi	Dickey's deer mouse	Tortuga I, Baja California
P. difficilis (*nasutus*)	Zacatecan deer mouse	Colorado – S Mexico; arid, rocky hills
P. eremicus (*collatus*)	Cactus mouse	Nevada – N Mexico
P. eva		Baja California
P. furvus (*latirostris*)	Blackish deer mouse	E Mexico; wet forest
P. gossypinus	Cotton mouse	SE USA; forest, swamps
P. grandis	Big deer mouse	Guatemala; montane forest
P. gratus		C Mexico
P. guardia	Angel Island mouse	Angel de la Guarda I, Baja California
P. guatemalensis (*altilaneus*)	Guatemalan deer mouse	S Mexico – Guatemala; montane forest

P. gymnotis (*allophylus*)		SE Mexico, Guatemala
P. hooperi	Hooper's mouse	Coahuila, N Mexico; ref. 9.13
P. interparietalis	San Lorenzo mouse	San Lorenzo Is, Baja California
P. leucopus	White-footed mouse	E, C USA – S Mexico; forest, scrub
P. madrensis	Tres Marias Island mouse	Tres Marias Is, W Mexico
P. maniculatus	Deer mouse	Labrador – Yukon – S Mexico
P. mayensis	Maya mouse	Guatemala; montane forest
P. megalops	Brown deer mouse	S Mexico; montane forest
P. mekisturus	Puebla deer mouse	Puebla, C Mexico; montane
P. melanocarpus	Zempoaltepec deer mouse	S Mexico; montane forest
P. melanophrys	Plateau mouse	C, S Mexico; steppe, desert
P. melanotis	Black-eared mouse	N, C Mexico; sibling of *P. maniculatus*
P. melanurus (*megalops*)	Black-tailed mouse	S Mexico
P. merriami	Merriam's mouse	NW Mexico, S Arizona; desert
P. mexicanus (*nudipes*)	Mexican deer mouse	E Mexico – W Panama; forest, scrub
P. ochraventer	El Carrizo deer mouse	NE Mexico; wet forest
P. oreas (*maniculatus*)		Washington – British Columbia
P. pectoralis	White-ankled mouse	Texas, N, C Mexico; desert, scrub
P. pembertoni	Pemberton's deer mouse	San Pedro Nolasco I, NW Mexico
P. perfulvus	Marsh mouse	WC Mexico
P. polionotus	Oldfield mouse	SE USA
P. polius	Chihuahuan mouse	Chihuahua, Mexico
P. pseudocrinitus	False canyon mouse	Coronados I, Baja California
P. sejugis	Santa Cruz mouse	Santa Cruz & San Diego Is, Baja California
P. simulus		NW Mexico
P. sitkensis	Sitka mouse	Alexander & Queen Charlotte Is, Alaska, W Canada
P. slevini	Slevin's mouse	Santa Catalina I, Baja California
P. spicilegus (*boylii*)		C Mexico

White-footed mouse
(Peromyscus leucopus)

Golden mouse
(Ochrotomys nuttalli)

P. stephani	San Esteban Island mouse	San Esteban I, Baja California
P. stirtoni	Stirton's deer mouse	Honduras, El Salvador, Guatemala; montane
P. truei (*comanche*)	Pinyon mouse	SW USA – S Mexico; scrub, steppe
P. winkelmanni	Winkelmann's mouse	Michoacan, W Mexico; montane
P. yucatanicus	Yucatan deer mouse	Yucatan, Mexico; forest
P. zarhynchus	Chiapan deer mouse	SE Mexico; montane forest

Habromys; (*Peromyscus*).

H. chinanteco		Oaxaco, Mexico
H. lepturus (*ixtlani*)		Oaxaco, Mexico; montane forest
H. lophurus	Crested-tailed mouse	SE Mexico – El Salvador; montane
H. simulatus	Jico deer mouse	NW Mexico; forest

Isthmomys; (*Peromyscus*); Panama, etc.; montane.

| *I. flavidus* | Yellow deer mouse | W Panama |
| *I. pirrensis* (*flavidus*) | Mount Pirri deer mouse | E Panama – NE Colombia |

Megadontomys; (*Peromyscus*).

| *M. thomasi* | Thomas's deer mouse | S Mexico; montane forest |

Osgoodomys; (*Peromyscus*).

| *O. banderanus* | Michoacan deer mouse | SW Mexico; dry forest |

Podomys; (*Peromyscus*).

| *P. floridanus* | Florida mouse | Florida |

Ochrotomys

| *O. nuttalli* | Golden mouse | SE USA; forest |

Baiomys; American pygmy mice; C, (N) America; grassland.

| *B. musculus* | Southern pygmy mouse | S Mexico – Nicaragua |
| *B. taylori* | Northern pygmy mouse | Texas, Arizona – C Mexico |

Onychomys; grasshopper mice; W North America; scrub, steppe.

O. arenicola		New Mexico, N Mexico
O. leucogaster	Northern grasshopper mouse	SW Canada – N Mexico
O. torridus	Southern grasshopper mouse	SW USA – C Mexico

Akodon; (*Abrothrix, Chroeomys, Hypsimys, Thaptomys*); South American field mice; S America; grassland, forest; classification very provisional; refs. 9.128, 129, 130

| *A. aerosus* (*urichi*) | | Peru — Bolivia |

A. affinis	Colombia
A. albiventer	Peru — N Argentina
A. andinus	C Chile — Peru; montane
A. azarae	NE Argentina — Bolivia — SE Brazil
A. boliviensis	S Peru — NW Argentina
A. budini	NW Argentina; montane
A. caenosus	NW Argentina, S Bolivia
A. cursor	Uruguay, Paraguay, SE Brazil
A. dayi	Bolivia
A. dolores	Sierra de Cordoba, C Argentina
A. fumeus	Bolivia, S Peru

South american field mouse
(*Akodon olivaceus*)

A. hershkovitzi	Islands W of Tierra del Fuego; ref. 9.14
A. illutens	NW Argentina
A. iniscatus	SC Argentina
A. jelskii	NW Argentina — C Peru; montane
A. kempi	Islands in Parana Estuary
A. kofordi	Peru; montane; ref. 9.130
A. lanosus	Patagonia
A. llanoi	I de los Estados, Tierra del Fuego; ref. 9.15
A. longipilis	Chile, W, C Argentina
A. mansoensis	Rio Negro, C Argentina; ref. 9.16
A. markhami	Wellington I, S Chile
A. molinae	E Argentina
A. mollis	Ecuador — Peru
A. neocenus	C & S Argentina
A. nigrita	E, S Brazil, N Argentina
A. nucus	SC Argentina
A. olivaceus	Chile, W Argentina
A. orophilus	Peru
A. pacificus	W Bolivia; montane
A. pervalens	Bolivia
A. puer	Bolivia, Peru; montane
A. reinhardti (*lasiotis*)	E Brazil
A. sanborni (*longipilis*)	Argentina, S Chile
A. serrensis	E, SE Brazil, N Argentina
A. siberiae	Bolivia; montane; ref. 9.129
A. simulator	Bolivia, Argentina
A. subfucus	Peru — N Argentina

A. surdus	SE Peru
A. sylvanus	Argentina
A. toba	W Paragua, E Bolivia, NW Argentina
A. tucumanensis	Peru — N Argentina
A. urichi	Peru — Venezuela, Trinidad; montane
A. varius	Bolivia
A. xanthorhinus	Patagonia, Tierra del Fuego

Bolomys; (*Akodon, Cabreramys*; *Thalpomys*); Peru – N Argentina; montane; ref. 9.79.

B. amoenus	SE Peru; montane
B. lactens	NW Argentina; montane
B. lasiurus (*lasiotis*) (*arviculoides*)	S, E Brazil
B. lenguarum (*tapirapoanus*)	Bolivia – Argentina
B. obscurus	Uruguay – C Argentina
B. temchuki	NE Argentina

Microxus; (*Akodon*).

M. bogotensis	Colombia, Venezuela
M. latebricola	Ecuador; montane
M. mimus	SE Peru; montane

Zygodontomys; cane mice; C, S America; savanna.

Z. borreroi	N Colombia
Z. brevicauda (*cherriei*) (*reigi*)	Costa Rica – Ecuador, Surinam, Trinidad

Podoxymys

P. roraimae	Roraima mouse	Guyana, etc.

Lenoxus

L. apicalis	Peruvian rat	Peru, Bolivia; montane forest

Oxymycterus; (*Akodon*); burrowing mice; S America S of Amazon; forest, cultivation.

O. akodontius	NW Argentina
O. angularis	E Brazil
O. delator	Paraguay
O. hiska	Peru; ref. 9.88
O. hispidus	NE Argentina – E Brazil
O. hucucha	Bolivia; ref. 9.88
O. iheringi	S Brazil, N Argentina
O. inca	W Bolivia – C Peru

O. paramensis		SE Peru – N Argentina
O. roberti		E Brazil
O. rufus		C, E Brazil –
(*rutilans*)		NE Argentina

Juscelinomys
J. candango — Brasilia, Brazil

Blarinomys
B. breviceps — Brazilian shrew-mouse — SE Brazil; montane forest

Notiomys
N. edwardsii — S Argentina

Geoxus; (*Notiomys*); ; predator; ref. 9.18.
G. valdivianus — Chile, S Argentina

Chelemys; (*Notiomys*); ref. 9.18.
C. macronyx — S Argentina, Chile
C. megalonyx — C Chile – Patagonia
 (*delfini*)

Kunsia; (*Scapteromys*).
K. fronto — SE Brazil, Paraguay, N Argentina
K. tomentosus — S Brazil, Bolivia; savanna, subterranean

Scapteromys
S. tumidus — Uruguay, NE Argentina, etc.; swamps

Bibimys; (*Akodon, Scapteromys*); ref. 9.17.
B. chacoensis — N, C Argentina
B. labiosus — SE Brazil
B. torresi — Parana delta, NE Argentina

Scotinomys; brown mice; central America; montane.
S. teguina — Alston's brown mouse — S Mexico – W Panama; montane forest
S. xerampelinus — Chiriqui brown mouse — Costa Rica, W Panama; montane forest, grass

Calomys; (*Baiomys, Hesperomys*); vesper mice; southern S America; grassland, scrub; ref. 9.86.
C. callidus — NE Argentina, S Paraguay
C. callosus — S Brazil – Bolivia – N Argentina; scrub
C. fecundus — Bolivia
C. hummelincki — Venezuela

C. laucha	C Argentina – Uruguay; grassland, scrub
C. lepidus	Peru, Bolivia, etc.; montane
C. musculinus	C Argentina – Paraguay
C. sorellus	Peru; montane

Eligmodontia; highland desert mice; Chile, etc.; arid scrub.

E. typus (*hypogaeus*) (*puerulus*)	S Peru – S Patagonia

Graomys; (*Phyllotis*).

G. domorum	N Argentina, Bolivia; montane
G. edithae	N Argentina
G. griseoflavus	S Argentina – Paraguay, Bolivia

Andalgalomys; (*Graomys*); ref. 9.77.

A. olrogi	NW Argentina; steppe
A. pearsoni	Paraguay – S Bolivia; grassland

Pseudoryzomys; (*Oryzomys*).

P. simplex	E Brazil
P. wavrini	Paraguay, N Argentina, Bolivia; marsh

Phyllotis; (*Galenomys, Paralomys*); leaf-eared mice; S America.

P. amicus	N Peru
P. andium	Peru, Ecuador; forest
P. bonaeriensis (*darwini*)	EC Argentina
P. caprinus	NW Argentina, S Bolivia, E of Andes; scrub
P. darwini	Patagonia – W Peru
P. definitus	Peru, W of Andes; forest
P. garleppi	Bolivia, S Peru
P. gerbillus	NW Peru; desert
P. haggardi	Ecuador
P. magister	SW Peru, N Chile; forest
P. osilae	S Peru – N Argentina; montane grassland
P. wolffsohni	W Bolivia

Leaf-eared mouse (Phyllotis)

Auliscomys; (*Phyllotis*); Andes; montane steppe.

A. boliviensis	S Peru – N Chile
A. micropus	Patagonia
A. pictus	Peru, W Bolivia
A. sublimus	NW Argentina – S Peru

Irenomys
I. tarsalis   Chilean rat   S Chile, S Argentina

Chinchillula
C. sahamae   Chinchilla mouse   NW Argentina – Peru

Punomys
P. lemminus   Puna mouse   Peru; high montane steppe

Neotomys
N. ebriosus   Andean swamp rat   Peru – NW Argentina; montane

Reithrodon
R. physodes   Pampas gerbil   Uruguay – Tierra del
(typicus)   (Rabbit rat)   Fuego; grassland

Euneomys; (*Chelemyscus*); Patagonian chinchilla-mice; Argentina, Chile; scrub, forest; ref. 9.89.
E. chinchilloides   Chile, W Argentina, Tierra
  (petersoni)   del Fuego
  (mordax)

Holochilus; marsh rats; S America; marshes.
H. brasiliensis   E Brazil – C Argentina
H. chacarius   Paraguay, NE Argentina
H. magnus   SE Brazil – NE Argentina
H. sciureus   C Brazil – Colombia –
  (brasiliensis)   Guianas

Sigmodon; cotton rats; S USA – Guyana; grassland, scrub.
S. alleni   Brown cotton rat   W, S Mexico
  (planifrons)
  (vulcani)
S. alstoni   Venezuela – Surinam
S. arizonae   Arizona cotton rat   Arizona, NW Mexico; E
S. fulviventer   Tawny-bellied cotton rat   SW Mexico – New Mexico
S. hispidus   Hispid cotton rat   S USA – Peru
S. leucotis   White-eared cotton rat   C Mexico; montane
  (alticola)
S. mascotensis   SW Mexico
S. ochrognathus   Yellow-nosed cotton rat   NC Mexico, S USA

Andinomys
A. edax   Andean mouse   Peru – NW Argentina, N Chile

Neotomodon
N. alstoni   Volcano mouse   C Mexico; montane

*Bushy-tailed wood rat
(Neotoma cinerea)*

Neotoma; woodrats; N, C America; desert – forest, especially rocky.

N. albigula (*latifrons*) (*montezumae*)	White-throated woodrat	SW USA, C Mexico
N. angustipalata	Tamaulipan woodrat	NE Mexico
N. anthonyi	Anthony's woodrat	Todos Santos I, Baja California
N. bryanti	Bryant's woodrat	Cerros I, Baja California
N. bunkeri	Bunker's woodrat	Coronados I, Baja California
N. chrysomelas	Nicaraguan woodrat	Nicaragua, Honduras; (in *N. mexicana*?)
N. cinerea	Bushy-tailed woodrat	S Yukon – Arizona
N. floridana	Eastern woodrat	SE USA; forest
N. fuscipes	Dusky-footed woodrat	Oregon – N Baja California; V
N. goldmani	Goldman's woodrat	NC Mexico
N. lepida (*devia*)	Desert woodrat	SW USA, Baja California; desert
N. martinensis	San Martin woodrat	San Martin I, Baja California
N. mexicana	Mexican woodrat	Colorado – Honduras
N. micropus	Southern plains woodrat	NE Mexico, W Texas, New Mexico, etc.
N. nelsoni	Nelson's woodrat	Veracruz, EC Mexico
N. palatina	Bolaños woodrat	Jalisco, W Mexico
N. phenax	Sonoran woodrat	NW Mexico; dry forest
N. stephensi	Stephen's woodrat	Arizona, W New Mexico
N. varia	Turner Island woodrat	Turner I, Gulf of California

Hodomys; (*Neotoma*).

H. alleni	Allen's woodrat	W, S Mexico; scrub

Nelsonia

N. neotomodon	Diminutive woodrat	WC Mexico; rocky forest

Xenomys

X. nelsoni	Magdalena rat	Jalisco, W Mexico; forest

Ichthyomys; fish-eating rats; NW South America; streams, swamps; ref. 9.90.

I. hydrobates		Ecuador, Colombia, W Venezuela
I. pittieri		Venezuela
I. stolzmanni		Ecuador, Peru
I. tweedii		Ecuador, Panama

Anotomys; (*Rheomys*); ref. 9.90.

A. leander	Ecuador fish-eating rat	Ecuador; montane streams

Chibchanomys; (*Anotomys*); ref. 9.90.

C. trichotis		Colombia, W Venezuela, Peru

Rheomys; water mice; C America; streams in forest; ref. 9.90.

R. mexicanus	Mexican water mouse	Oaxaca, Mexico
R. raptor (*hartmanni*)	Goldman's water mouse	Panama, Costa Rica
R. thomasi	Thomas' water mouse	El Salvador, Chiapas
R. underwoodi	Underwood's water mouse	Costa Rica, W Panama

Neusticomys; (*Daptomys*); ref. 9.90.

N. monticolus	Fish-eating mouse	Ecuador, Colombia; montane
N. oyapocki		French Guiana
N. peruviensis	Peruvian fish-eating rat	EC Peru; streams in lowland forest
N. venezuelae	Venezuelan fish-eating rat	N Venezuela, Guyana

Subfamily Cricetinae

Hamsters etc.; *c.* 25 species; Europe, N Asia; grasslands, steppe; seed-eaters.

Calomyscus; mouse-like hamsters; species similar, sometimes all included in *C. bailwardi*.

Common hamster
(*Cricetus cricetus*)

C. bailwardi	Iran, Afghanistan, etc.
C. baluchi	E Afghanistan, W Pakistan
C. hotsoni	W Pakistan
C. mystax	S Turkmenia, etc.
C. urartensis	NW Iran, etc.

Phodopus

P. campbelli (*sungorus*)		Mongolia, etc. (in *P. sungorus*?)
P. roborovskii	Desert hamster	Mongolia – Shaanxi
P. sungorus	Striped hairy-footed hamster	E Kazakhstan, SW Siberia

Cricetus

C. cricetus	Common hamster	W Europe – Altai; grassland

Cricetulus; (*Allocricetulus, Tscherskia*); dwarf hamsters; SE Europe – China.

C. alticola	Ladak hamster	Ladak, Kashmir
C. barabensis	Striped hamster	S Siberia – Manchuria, Korea
C. curtatus	Mongolian hamster	Mongolia
C. eversmanni	Eversmann's hamster	N Kazakhstan
C. griseus (*barabensis*)		NE China

C. kamensis	Tibetan hamster	Tibetan plateau
C. longicaudatus	Lesser long-tailed hamster	NW China, Mongolia, etc.
C. migratorius	Grey hamster	SE Europe – Sinkiang
C. obscurus (*barabensis*)		Mongolia, NW China
C. pseudogriseus (*barabensis*)		N Mongolia, etc.
C. sokolovi		Mongolia, W China; ref. 9.92
C. triton	Greater long-tailed hamster	NE China, Korea, Ussuri

Cansumys; (*Cricetulus*, *Tscherskia*).

C. canus	Gansu hamster	Gansu, China; ref. 9.93

Mesocricetus

M. auratus (*brandti*)	Golden hamster	Asia Minor, etc.
M. newtoni	Rumanian hamster	E Rumania, Bulgaria
M. raddei	Ciscaucasian hamster	Steppes N of Caucasus

Subfamily Spalacinae

Blind mole-rats; *c.* 8 species; E Mediterranean – S Russia; grassland, cultivation; subterranean; vegetarian.

Blind mole-rat
(Nannospalax ehrenbergi)

Spalax

S. arenarius	S Ukraine
S. giganteus	N of Caspian Sea
S. graecus	N Rumania, Ukraine
S. microphthalmus	S Russia
S. polonicus	W Ukraine

Nannospalax; (*Spalax*, *Microspalax*, *Mesospalax*).

N. ehrenbergi (*leucodon*)	Syria – Libya
N. leucodon	Yugoslavia, Greece – SW Ukraine
N. nehringi (*leucodon*)	Asia Minor

Subfamily Myospalacinae

Zokors (eastern Asiatic mole-rats); *c.* 6 species; Altai – China; grassland, steppe.

Myospalax

M. aspalax (*myospalax*)		Transbaikalia, N Mongolia
M. fontanierii (*baileyi*) (*cansus*)	Common Chinese zokor	Sichuan – Hebei
M. myospalax	Siberian zokor	Altai, W Siberia

Zokor
(Nyospalax aspalax)

M. psilurus *(myospalax)*		N China, Transbaikalia
M. rothschildi	Rothschild's zokor	Gansu, Hubei
M. smithii	Smith's zokor	Gansu

Crested rat
(Lophiomys imhaustii)

Subfamily Lophiomyinae

Lophiomys

L. imhausii	Crested rat	E Africa; montane forest

Subfamily Platacanthomyinae

Spiny dormice; 2 species; SE Asia; forest.

Platacanthomys

P. lasiurus	Malabar spiny dormouse	S India; forest, rocks

Typhlomys

T. cinereus	Chinese pygmy dormouse	SE China, N Indochina; montane forest

Spiny dormouse
(Platacanthomys lasiurus)

Subfamily Nesomyinae

Madagascan rats, etc.; *c.* 11 species; Madagascar, (S Africa); forest, grassland.

Macrotarsomys

M. bastardi	W Madagascar
M. ingens	NW Madagascar; forest

Nesomys

N. rufus	Madagascar

Madagascan rat
(Macrotarsomys bastardi)

Brachytarsomys

B. albicauda	E Madagascar; forest

Eliurus

E. minor	E Madagascar; forest
E. myoxinus	Madagascar; forest

Gymnuromys

G. roberti	E Madagascar

Hypogeomys

H. antimena	W Madagascar; sandy coastal forest

Brachyuromys

B. betsileoensis	Madagascar
B. ramirohitra	E Madagascar

Mystromys

M. albicaudatus	White-tailed rat	S Africa

Vlei rat
(Otomys irroratus)

Subfamily Otomyinae

African swamp rats; *c.* 13 species; Africa S of Sahara; grassland, swamps.

Otomys; (*Myotomys*).

O. anchietae		Angola, S Tanzania
O. angoniensis	Angoni vlei rat	S Africa – Angola – Kenya
O. denti		NE Zambia – Uganda; montane
O. irroratus	Vlei rat	S Africa – Zimbabwe
O. laminatus	Laminate vlei rat	S Africa
O. maximus	Large vlei rat	SW Zambia, etc.
O. saundersiae	Saunder's vlei rat	Cape Province, etc.
O. sloggetti	Sloggett's vlei rat	S Africa; montane
O. tropicalis		Cameroun – Kenya
O. typus		Zambia – Ethiopia; montane
O. unisulcatus	Bush Karroo rat	Cape Province

Parotomys; Karroo rats.

P. brantsii	Brant's whistling rat	Cape Province, etc.
P. littledalei	Littledale's whistling rat	Cape Province, Namibia

Subfamily Rhizomyinae

East African mole-rats, bamboo rats; *c.* 6 species; E Africa, SE Asia; montane forest, grassland, subterranean vegetarians.

East African mole rat
(Tachyoryctes splendeus)

Tachyoryctes

T. macrocephalus	Giant mole-rat	Ethiopia; montane grassland
T. splendens	East African mole-rat	E Africa; montane

Rhizomys; bamboo rats; SE Asia.

R. pruinosus	Hoary bamboo rat	Assam, S China – Malaya
R. sinensis	Chinese bamboo rat	S China, N Burma
R. sumatrensis	Large bamboo rat	Indochina – Malaya, Sumatra

Cannomys

C. badius	Lesser bamboo rat (Bay bamboo rat)	Nepal – Thailand

Norway lemming
(Lemmus lemmus)

Subfamily Arvicolinae (Microtinae).

Voles, lemmings; *c.* 130 species; N America, N Eurasia; tundra, grassland, scrub, open forest; herbivores.

Dicrostonyx; collared lemmings; tundra; classification very provisional.

D. exsul	St Lawrence Island collared lemming	St Lawrence I (Alaska)

D. groenlandicus (*torquatus*)	American arctic lemming	Greenland, N Canada (Arctic Islands)
D. hudsonius	Labrador collared lemming	Labrador, N Quebec
D. nelsoni	Nelson's collared lemming	Alaska
D. richardsoni	Richardson's collared lemming	Canada W of Hudson Bay
D. rubricatus	Bering collared lemming	N Alaska
D. stevensoni	Stevenson's collared lemming	Umnak I (Alaska)
D. torquatus	Siberian arctic lemming (Collared lemming)	Siberia
D. vinogradovi	Wrangel lemming	Wrangel I, Siberia

Synaptomys; bog lemmings; N America.

S. borealis	Northern bog lemming	Canada, Alaska
S. cooperi	Southern bog lemming	NE USA, SE Canada

Myopus; (*Lemmus*).

M. schisticolor	Wood lemming	Scandinavia, Siberia; taiga

Lemmus; brown lemmings; tundra.

L. amurensis	Amur lemming	Verkhoyansk Mts – Upper Amur
L. chrysogaster (*sibiricus*)		NE Siberia; (in *L. sibiricus* ?)
L. lemmus	Norway lemming	Scandinavia
L. nigripes	Black-footed lemming	Pribilof Is
L. sibiricus (*trimucronatus*)	Siberian lemming (Brown lemming)	N Siberia W to White Sea; N, W Canada, Alaska

Clethrionomys; (*Evotomys, Eothenomys*); red-backed voles, bank voles; forest, scrub, tundra.

C. andersoni	Japanese red-backed vole	Honshu, Japan
C. californicus (*occidentalis*)	Western red-backed vole	Oregon – N California
C. centralis (*frater*)		Tien Shan Mts, etc.
C. gapperi	Gapper's red-backed vole	Canada (except NW), N USA, Rocky Mts, Appalachians; forest
C. glareolus	Bank vole	Europe – Altai; dec. forest
C. rufocanus (*rex*)	Grey red-backed vole	Scandinavia, Siberia, Hokkaido; tundra, montane
C. rutilus	Northern red-backed vole	N Eurasia, Arctic America; tundra, taiga

Eothenomys; (*Aschizomys*); oriental voles; E Asia.

E. chinensis		S China; montane
E. custos		S China; montane

E. eva		W China; montane
E. inez		C China
E. lemminus		NE Siberia; tundra
E. melanogaster	Père David's vole	S China, Taiwan
E. olitor		Yunnan, S China; montane
E. proditor		S China; montane
E. regulus		Korea, E Manchuria
E. shanseius		Shanxi, China
E. smithii		Japan, except Hokkaido

Alticola; mountain voles; C Asia; ref. 9.95.

A. barakshin (*stoliczkanus*)		W, SW Mongolia
A. fetisovi		E Siberia; ref. 9.94
A. macrotis	Large-eared vole	Altai, Sayan Mts
A. roylei	Royle's mountain vole	W Himalayas – Altai; montane
A. semicanus (*roylei*)		Mongolia
A. stoliczkanus (*stracheyi*)	Stoliczka's mountain vole	Himalayas – Altai
A. strelzowi	Flat-headed vole	Altai, E Kazakhstan
A. tuvinicus (*roylei*)		Altai Mts – L Baikal

Hyperacrius

| *H. fertilis* | True's vole | Kashmir, N Punjab |
| *H. wynnei* | Murree vole | N Pakistan; montane forest |

Dinaromys; (*Dolomys*).

| *D. bogdanovi* | Martino's snow vole | Yugoslavia; montane |

Arvicola; water voles; N Eurasia; freshwater banks, grassland.

| *A. sapidus* | Southwestern water vole | Iberia, SW France |
| *A. terrestris* (*amphibius*) | European water vole | Europe – E Siberia |

Ondatra

| *O. zibethicus* | Muskrat | N America, [N Eurasia]; freshwater banks |

Neofiber

| *N. alleni* | Round-tailed muskrat (Florida water rat) | Florida; freshwater banks, marshes |

Phenacomys; (*Arborimus*)

| *P. albipes* | White-footed vole | W Oregon, NW California; coastal forest |

P. intermedius	Heather vole	Canada, W USA; forest, tundra
P. longicaudus (*silvicola*)	Red tree vole	Oregon, NW California

Pitymys; (*Blanfordimys*, *Microtus*, *Neodon*); pine voles; N America, Eurasia; doubtfully distinct from *Microtus*.

P. afghanus	Afghan vole	Afghanistan, S Turkestan; montane
P. bavaricus†	Bavarian pine vole	S Germany; montane; Ex
P. daghestanicus	Daghestan pine vole	E Caucasus
P. duodecimcostatus	Mediterranean pine vole	S, E Iberia – SE France
P. felteni (*savii*)		S Yugoslavia
P. gerbii (*savii*)		SW France, NE Spain
P. guatemalensis	Guatemalan vole	Guatemala; montane
P. juldaschi	Juniper vole	Tien Shan, Pamirs; montane
P. leucurus	Blyth's vole	Tibetan Plateau, Himalayas
P. liechtensteini	Liechtenstein's pine vole	NW Yugoslavia
P. lusitanicus	Lusitanian pine vole	NW Iberia, SW France
P. majori	Major's pine vole	Caucasus, Asia Minor
P. multiplex	Alpine pine vole	European Alps
P. ochrogaster	Prairie vole	C, S USA
P. pinetorum	American pine vole	E USA; forest
P. quasiater	Jalapan pine vole	EC Mexico
P. savii	Savi's pine vole	Italy, S France
P. schelkovnikovi	Schelkovnikov's pine vole	Elburz, Talysh Mts, S of Caspian Sea
P. sikimensis	Sikkim vole	Himalayas – W China; montane
P. subterraneus	European pine vole	France – C Russia
P. tatricus	Tatra pine vole	Tatra Mts; montane
P. thomasi	Thomas' pine vole	S Balkans

Microtus; grass voles, meadow voles; N America, N Eurasia, (N Africa); grassland, open forest, tundra.

M. abbreviatus	Insular vole	St Matthew I, Hall I, Bering Sea; vicariant of *M. gregalis*
M. agrestis	Field vole	Europe – R Lena, E Siberia
M. arvalis	Common vole	Europe – R Yenesei; grassland
M. bedfordi	Duke of Bedford's vole	Gansu, China
M. brandtii	Brandt's vole	Mongolia, Transbaikalia
M. cabrerae	Cabrera's vole	Iberia

Field vole
(Microtus agrestis)

M. californicus	California vole	California; (E)
M. canicaudus	Grey-tailed vole	Oregon; close to *M. montanus*
M. chrotorrhinus	Rock vole	E Canada, NE USA
M. clarkei	Clarke's vole	Yunnan, N Burma; high montane
M. evoronensis		Lower Amur, E Siberia; ref. 9.20
M. fortis	Reed vole	China, SE Siberia
M. gregalis	Narrow-headed vole	Siberia, C Asia
M. gud		NE Asia Minor, Caucasus
M. guentheri	Gunther's vole	SE Europe – Israel, Libya
M. irani	Persian vole	Iran, etc.
M. kermanensis	Baluchistan vole	SE Iran; ref. 9.96
M. kikuchii	Taiwan vole	Taiwan; montane
M. kirgisorum	Tien Shan vole	Tien Shan Mts
M. limnophilus (*oeconomus*)		W Mongolia, W China
M. longicaudus (*coronarius*)	Long-tailed vole	SW USA – Alaska
M. mandarinus	Mandarin vole	C China – SE Siberia
M. maximowiczii (*ungurensis*)		Upper Amur, E Siberia
M. mexicanus (*fulviventer*)	Mexican vole	Mexico, S USA; montane
M. middendorffii (*hyperboreus*)	Middendorff's vole	N Siberia; tundra
M. millicens	Szechwan vole	Sichuan, China
M. miurus	Singing vole	Alaska, Yukon
M. mongolicus	Mongolian vole	NE Mongolia, etc.
M. montanus	Montane vole	W USA; montane
M. montebelli	Japanese grass vole	Japan
M. mujanensis		Vitim Basin, E Siberia; ref. 9.78
M. nivalis	Snow vole	SW Europe – Iran; montane
M. oaxacensis	Tarabundi vole	Oaxaca, Mexico; montane forest
M. oeconomus	Root vole	N Eurasia, Alaska, etc.; tundra, taiga
M. oregoni	Creeping vole	W coast USA
M. pennsylvanicus (*breweri*)	Meadow vole	Canada, N USA; (R)
M. richardsoni	American water vole	NW USA, SW Canada; montane
M. roberti	Robert's vole	NE Asia Minor, W Caucasus; forest

M. rossiaemeridionalis   Finland, Greece – Urals,
  (epiroticus)   Caucasus; close to
  (subarvalis)   *M. arvalis*; ref. 9.21
  (arvalis)
M. sachalinensis   Sakhalin vole   Sakhalin I
M. socialis   Social vole   Kazakhstan – Palestine
M. townsendii   Townsend's vole   W coast USA, Vancouver I
M. transcaspicus   Transcaspian vole   SW Turkestan; steppe
M. umbrosus   Zempoaltepec vole   Oaxaca, Mexico; montane
M. xanthognathus   Yellow-cheeked vole   NW Canada, Alaska;
                                          forest, tundra

Lemmiscus; (*Lagurus*).
L. curtatus   Sagebrush vole   W USA; montane steppe

Lagurus
L. lagurus   Steppe lemming   Ukraine – Mongolia,
                                Sinkiang

Eolagurus; (*Lagurus*); yellow steppe lemmings.
E. luteus   W Mongolia, N Sinkiang,
                                  († Kazakhstan)
E. przewalskii   N Tibet – S Mongolia
  (lutens)

Prometheomys
P. schaposchnikowi   Long-clawed mole-vole   Caucasus, NE Asia Minor

Ellobius; mole-voles; C Asia; steppe.
E. alaicus   Kirghizia
  (talpinus)
E. fuscocapillus   Southern mole-vole   SW Turkestan –
                                            Baluchistan
E. lutescens   Caucasus – W Iran, etc.
  (fuscocapillus)
E. talpinus   Northern mole-vole   Ukraine – Kazakhstan
E. tancrei   Uzbekistan – Sinkiang;
                                  ref. 9.97

Subfamily Gerbillinae

Gerbils, jirds; *c.* 97 species; C, W Asia, Africa; steppe, desert; seed-eaters.

Gerbillus; (*Dipodillus, Monodia*); northern pygmy gerbils; N, E Africa, SW Asia;
desert, steppe; classification very provisional; ref. 9.22.

G. acticola   Somalia
G. agag   Sudan, ? Chad
  (? dalloni)
G. allenbyi   Israel; coastal dunes
G. amoenus   Egypt – Mauritania
  (nanus)

*Gerbil
(Gerbillus gerbillus)*

G. andersoni (? *bonhotei*)	Anderson's gerbil	Israel – Tunisia
G. bottai		Sudan, Kenya
G. burtoni (*pyramidum*)		Sudan
G. campestris (? *lowei*)	Large North African gerbil	Morocco – Egypt, Sudan
G. cheesmani (*aquilus*)	Cheesman's gerbil	Arabia – SW Iran
G. cosensi (*agag*)		N Kenya
G. dasyurus	Wagner's gerbil	Arabia, Israel, Iraq
G. dunni		Ethiopia, Somalia
G. famulus	Black-tufted gerbil	SW Arabia
G. garamantis (*nanus*)		Algeria
G. gerbillus (*hirtipes*)		Morocco – Israel
G. gleadowi	Indian hairy-footed gerbil	NW India, Pakistan
G. grobbeni (*nanus*)		NE Libya
G. harwoodi		Kenya
G. henleyi	Pygmy gerbil	Algeria – Israel – Oman
G. hesperinus		Morocco
G. hoogstraali		Morocco
G. jamesi		Tunisia
G. latastei (*gerbillus*) (*pyramidum*)		Libya, Tunisia, ? Algeria
G. mackilligini (*nanus*)		S Egypt, ? N Sudan
G. maghrebi	Greater short-tailed gerbil	Morocco
G. mauritaniae		Mauritania
G. mesopotamiae	Mesopotamian gerbil	Iraq, SW Iran
G. muriculus		Sudan
G. nancillus		Sudan
G. nanus (? *brockmani*)	Baluchistan gerbil	Morocco – Israel – W Pakistan; ? Somalia
G. nigeriae (*agag*)		N Nigeria
G. occiduus		Morocco
G. perpallidus		N Egypt
G. poecilops	Large Aden gerbil	Arabia
G. principulus (*nanus*)		Sudan
G. pulvinatus (? *bilensis*)		Ethiopia
G. pusillus (? *diminutus*) (? *percivali*)		Kenya, Ethiopia

G. pyramidum (? *dongolanus*) (? *floweri*)	Greater Egyptian gerbil	Egypt, Sudan
G. riggenbachi		Rio de Oro, W Sahara
G. rosalinda		Sudan
G. ruberrimus		Somalia, Kenya
G. simoni (*kaiseri*) (*zakariai*)	Lesser short-tailed gerbil	Algeria – Egypt
G. somalicus (*campestris*)		Somalia
G. stigmonyx (*campestris*)		Sudan
G. syrticus		Libya
G. tarabuli (*pyramidum*)		Libya
G. vivax (*nanus*)		Libya
G. watersi (? *juliani*)		Somalia, Sudan

Gerbillurus; (*Gerbillus*); southern pygmy gerbils; SW Africa; desert, subdesert.

G. paeba	Hairy-footed gerbil	SW Angola – Cape Province; desert, subdesert
G. setzeri	Setzer's hairy-footed gerbil	Namib Desert; gravel plains
G. tytonis	Dune hairy-footed gerbil	S Namib, SW Africa; desert
G. vallinus	Bushy-tailed hairy-footed gerbil	SW Angola – Orange R

Microdillus

M. peeli	Somalia pygmy gerbil	Somalia; semidesert

Tatera; rat-like gerbils; Africa, (SW Asia); steppe, savanna; ref. 9.98.

T. afra	Cape gerbil	SW Cape Province; steppe
T. boehmi	Boehm's gerbil	Uganda – Zambezi, Angola
T. brantsii	Highveld gerbil	S Africa – Zambia
T. guineae (*robusta*)		W Africa
T. inclusa	Gorongoza gerbil	E Zimbabwe, Mozambique, E Tanzania
T. indica	Indian gerbil	W India – Asia Minor, Sri Lanka
T. leucogaster	Bushveld gerbil	S, C Africa; savanna
T. nigricauda	Black-tailed gerbil	S Somalia, Kenya, N Tanzania

T. phillipsi (*robusta*)		Kenya, Ethiopia, Somalia
T. robusta	Fringe-tailed gerbil	Tanzania – Sudan; Burkina Faso
T. valida (*gambiana*)	Savanna gerbil	Senegal – Ethiopia – Angola; savanna

Taterillus; W, C, E Africa; savanna.

T. arenarius		Mauritania, Niger, Mali; sahel
T. congicus		Cameroun – Sudan
T. emini		Chad – Kenya, Ethiopia
T. gracilis		W Africa; guinea savanna
T. harringtoni		Sudan – NE Tanzania; dry savanna
T. lacustris		L Chad, etc.
T. petteri		Burkina Faso; ref. 9.99
T. pygargus		Senegal – Mali

Desmodillus

D. auricularis	Short-tailed gerbil (Cape short-eared gerbil)	S Africa

Desmodilliscus

D. braueri	Pouched gerbil	Sudan – Senegal

Pachyuromys

P. duprasi	Fat-tailed gerbil	N Sahara

Ammodillus

A. imbellis	Somali gerbil	Somalia, SE Ethiopia

Sekeetamys; (*Meriones*).

S. calurus	Bushy-tailed jird	E Egypt, S Israel – C Arabia

Mongolian gerbil
(Meriones unguicalatus)

Meriones; jirds, gerbils; Sahara – China; steppe, desert.

M. chengi		Turfan, Sinkiang
M. crassus		Sahara – Afghanistan; desert
M. dahli		Armenia; ref. 9.100
M. hurrianae	Indian desert gerbil	NW India – SE Iran
M. libycus (*caudatus*)	Libyan jird	Sahara – Sinkiang; desert
M. meridianus	Mid-day gerbil	Russian Turkestan – China
M. nogaiorum		N of Caucasus; ref. 9.100
M. persicus	Persian jird	Asia Minor – Pakistan
M. rex	King jird	SW Arabia
M. sacramenti	Buxton's jird	S Israel
M. shawii	Shaw's jird	Morocco – Egypt

M. tamariscinus	Tamarisk gerbil	Lower Volga – W Gansu
M. tristrami	Tristram's jird	Sinai – Asia Minor, NW Iran
M. unguiculatus	Mongolian gerbil (Clawed jird)	Mongolia, Sinkiang – Manchuria
M. vinogradovi	Vinogradov's jird	E Asia Minor – N Iran
M. zarudnyi		NE Iran, etc.

Brachiones
B. przewalskii	Przewalski's gerbil	Sinkiang – Gansu

Psammomys
P. obesus	Fat sand rat	Algeria – Arabia
P. vexillaris		Algeria – Libya

Rhombomys
R. opimus	Great gerbil	Caspian Sea – Sinkiang, Pakistan

Subfamily Dendromurinae

African climbing mice, etc.; *c.* 22 species; Africa S of Sahara; forest – grassland.

Delanymys
D. brooksi		SW Uganda, etc.; montane swamps

Large-eared mouse (Malacothrix typica)

Dendromus; African climbing mice; Africa S of Sahara; forest – grassland.
D. kahuziensis		E Zaire; montane
D. lovati		Ethiopia; montane
D. melanotis	Grey climbing mouse	S Africa – Ethiopia – Guinea; savanna
D. mesomelas	Brants's climbing mouse	S Africa – Ethiopia – Cameroun
D. messorius (*mystacalis*)		Nigeria – Zaire
D. mystacalis	Chestnut climbing mouse	S Africa – Ethiopia – Zaire
D. nyikae	Nyika climbing mouse	E Transvaal – Malawi – Angola

Dendroprionomys
D. rousseloti		Congo Republic; forest

Deomys
D. ferrugineus	Congo forest mouse	Zaire, Cameroun – Uganda; rain forest

Leimacomys
L. buettneri		Togo; forest

Malacothrix
M. typica	Large-eared mouse (Gerbil-mouse)	S Africa – S Angola

Megadendromus

M. nikolausi		E Ethiopia; montane; ref. 9.23

Petromyscus; rock mouse; SW Africa; rocky steppe.

P. collinus	Pygmy rock mouse	SW Cape Province – S Angola
P. monticularis	Brukkaros rock mouse	SW Africa

Steatomys; fat mice; Africa S of Sahara; savanna.

S. caurinus		Senegal – Nigeria
S. cuppedius		Senegal – Niger
S. jacksoni		Ghana, W Nigeria
S. krebsii	Krebs' fat mouse	S Africa – Angola, Zambia
S. parvus (*minutus*)	Tiny fat mouse	Senegal – Somalia – SW Africa; dry savanna
S. pratensis (*caurinus*)	Common fat mouse	Senegal – Sudan – Natal; savanna, dry forest

Prionomys

P. batesi	Dollman's tree mouse	Cameroun, Cent. African Rep.

Subfamily Cricetomyinae

African pouched rats; *c.* 5 species; Africa S of Sahara; savanna, forest.

Giant pouched rat
(Cricetomys gambianus)

Beamys; long-tailed pouched rats.

B. hindei (*major*)		S Kenya, Malawi; NE Zambia

Saccostomus; short-tailed pouched rats; savanna; ref. 9.24.

S. campestris	Pouched mouse	S Africa – Zambia
S. mearnsi		Tanzania – Ethiopia, etc.

Cricetomys; giant pouched rats.

C. emini		Sierra Leone – L Tanganyika; rain forest
C. gambianus		Senegal – Sudan – S Africa; savanna

Subfamily Murinae

Old world mice and rats; *c.* 457 species; Eurasia, Africa, Australasia; forest (grassland); mainly seed-eaters.

Hapalomys; marmoset-rats; SE Asia; forest.

H. delacouri		Indochina, Hainan
H. longicaudatus		S Burma, Malaya

Harvest mouse
(Micromys minutus)

Vernaya; oriental climbing mice; S China, etc.; montane forest.

V. foramena	Sichuan climbing mouse	N Sichuan, China; ref. 9.25
V. fulva	Vernay's climbing mouse	S China, N Burma; montane forest

Tokudaia

T. osimensis	Ryukyu spiny rat	Ryukyu Is; forest

Vandeleuria

V. nolthenii		Sri Lanka; montane
V. oleracea	Palm mouse	India – Indochina, Sri Lanka; forest

Micromys

M. minutus (*danubialis*)	Harvest mouse	N Eurasia; grassland

Apodemus; (*Sylvaemus*); wood mice (long-tailed field mice); Palaearctic; forest, (grassland).

A. agrarius	Striped field mouse	C Europe – China
A. alpicola		European Alps; ref. 9.127
A. argenteus	Small Japanese field mouse	Japan
A. chevrieri (*agrarius*)		SW China; montane
A. draco (*semotus*)		S China, etc., Taiwan
A. falzfeini		Crimea etc.; ref. 9.122
A. flavicollis	Yellow-necked mouse	Europe, Asia Minor
A. gurkha	Himalayan field mouse	Nepal; montane forest
A. latronum		SW China, N Burma
A. microps	Pygmy field mouse	E Europe, Asia Minor; grassland, scrub
A. mystacinus	Broad-toothed mouse	SE Europe, Asia Minor, etc., scrub, rocks
A. peninsulae	Korean field mouse	Manchuria, Korea, etc., Hokkaido
A. speciosus	Large Japanese field mouse	Japan
A. sylvaticus (*krkensis*)	Wood mouse	Europe – Altai, Himalayas; N Africa

Thamnomys

T. rutilans		W Africa – Uganda, N Angola
T. venustus		E Zaire, etc.; montane forest

Grammomys; (*Thamnomys*); Africa S of Sahara; ref. 9.101.

G. aridulus		Sudan
G. buntingi (*dolichurus*)		Sierra Leone – Ivory Coast

G. caniceps		Kenya coast
G. cometes		S Africa – Sudan
G. dolichurus	Woodland mouse	S Africa – Cent. Afr. Rep.; dry forest
G. gigas (*dolichurus*)		Mt Kenya
G. macmillani		Ethiopia – Zaire
G. minnae		Ethiopia

Carpomys; Luzon rats; Philippines; montane forest.

C. melanurus		Luzon, Philippines
C. phaeurus		Luzon, Philippines

Batomys; (*Mindanaomys*); Philippine forest rats; montane forest.

B. dentatus		Luzon, Philippines
B. granti		Luzon, Philippines; I
B. salomonseni		Mindanao etc., Philippines

Pithecheir

P. melanurus		Java; montane forest
P. parvus	Malayan tree rat (Monkey-footed rat)	Malaya

Hyomys

H. goliath	Rough-tailed giant rat (White-eared giant rat)	New Guinea

Conilurus

C. albipes	White-footed rabbit-rat	E, S Australia; probably extinct
C. penicillatus	Brush-tailed rabbit-rat	N, NW Australia, New Guinea; savanna

Zyzomys; (*Laomys*); Australian rock rats; N, C Australia; rocky outcrops in desert to savanna; ref. 9.124.

Z. argurus	Common Australian rock rat	W, N Australia
Z. maini		N Australia
Z. palatalis		N Australia
Z. pedunculatus	Macdonnell Range rock rat (Central rock rat)	C Australia; I
Z. woodwardi	Woodward's rock rat (Large rock rat)	NW Australia

Mesembriomys

M. gouldii	Black-footed tree rat	N Australia; savanna
M. macrurus	Golden-backed tree rat	NW, N Australia; savanna

Oenomys

O. hypoxanthus	Rufous-nosed rat	Sierra Leone – N Angola, E Africa; forest

Lamottemys; ref. 9.105.

L. okuensis   Cameroun

Mylomys; (*Pelomys*).

M. dybowskyii   Ivory Coast – Kenya,
                                                        Zaire; savanna

Dasymys

D. incomtus   African marsh rat   Senegal – Ethiopia –
                              (African water rat)   S Africa; marshes, river
                                                                      banks

Arvicanthis; African grass rats; Africa, Arabia; savanna; taxonomy very
provisional.

A. abyssinicus   W Africa – Ethiopia –
                                                        Zambia

A. blicki   Ethiopia; high montane
A. dembeensis   Ethiopia – ? W Africa;
    (*testicularis*)   ? SW Arabia
    (? *naso*)
A. niloticus   Egypt
A. somalicus   Somalia, Ethiopia, Kenya

Hadromys

H. humei   Manipur bush rat   Assam – Yunnan; open
                                                            montane forest

Golunda

G. ellioti   Indian bush rat   India, etc., Sri Lanka;
                                                              marsh, cultivation

Pelomys; (*Desmomys*); groove-toothed swamp rats; Africa.

P. campanae   Zaire, W Angola
P. fallax   Creek rat   Kenya – Botswana
P. harringtoni   C Ethiopia; montane
P. hopkinsi   Rwanda, SW Uganda;
                                                          marshes
P. isseli   Kome & Bussu Is,
                                                          L Victoria
P. minor   Zaire, Angola, Zambia
P. rex   Ethiopia

Lemniscomys; striped grass rats; Africa; savanna.

L. barbarus   Senegal – Ethiopia,
                                                          Tanzania, NW Africa;
                                                          dry savanna, steppe
L. bellieri   Ivory Coast; savanna;
                                                          ref. 9.26
L. griselda   Single-striped mouse   Angola
L. linulus   Senegal – N Ivory Coast;
    (*griselda*)   close to *L. rosalia*

Striped grass rat
(Lemniscomys striatus)

L. macculus		Uganda, etc.
L. mittendorfi (*striatus*)		Cameroun
L. rosalia (*griselda*)		Kenya – S Africa
L. roseveari		Zambia; ref. 9.27
L. striatus		Sierra Leone – Ethiopia – Malawi – Angola; savanna

Rhabdomys

R. pumilio	Four-striped grass mouse	S Africa – Uganda, Kenya; grassland

Hybomys; (*Typomys*); tropical Africa; forest; refs. 9.102, 103, 104.

H. basilii		Bioko I
H. eisentrauti		W Cameroun; montane
H. lunaris (*univittatus*)		Ruwenzori Mts
H. planifrons (*trivirgatus*)		Sierra Leone, Liberia
H. trivirgatus		Ivory Coast – Nigeria
H. univittatus		E Nigeria – Uganda

Millardia; Indian soft-furred rats; India, etc.; grassland, cultivation.

M. gleadowi	Sand-coloured rat	Pakistan, NW India
M. kathleenae		Burma
M. kondana		W India; ref. 9.28
M. meltada	Soft-furred field rat	India, Sri Lanka

Dacnomys

D. millardi	Millard's rat	Assam – Indochina; montane forest

Eropeplus

E. canus	Sulawesi soft-furred rat	Sulawesi

Stenocephalemys; Ethiopian narrow-headed rats; Ethiopia; high montane; should perhaps be included in *Praomys*.

S. albocaudata		Ethiopia
S. griseicauda		Ethiopia

Rattus; (*Nesoromys*); Old-world rats; SE Asia, Australia, some worldwide by association with man; mainly forest, scrub, cultivation; taxonomy still very provisional.

R. adustus		Enganno I (Sumatra); ref. 9.106
R. annandalei	Annandale's rat	S Thailand, Malaya, Sumatra

Brown rat
(Rattus norvegicus)

R. argentiventer	Ricefield rat	Indochina – Java, Borneo, Philippines, [Sulawesi, New Guinea]; cultivation, scrub
R. baluensis	Summit rat	Borneo; montane forest
R. ceramicus	Ceram rat	Ceram, Indonesia; montane forest
R. doboensis		Aru Is, Indonesia
R. elaphinus		Taliabu, Sulawesi
R. enganus		Enggano I (Sumatra)
R. everetti		Philippines
R. exulans (*bocourti*) (*ephippium*)	Polynesian rat	Bangladesh – New Guinea, Philippines, New Zealand, Pacific islands; commensal
R. feliceus (*ruber*)		Ceram, Indonesia
R. fuscipes	Bush rat	E, S, SW Australia; coastal
R. giluwensis		C New Guinea; ref. 9.33
R. hoffmanni (*tatei*)		Sulawesi
R. hoogerwerfi		N Sumatra; montane
R. hoxaensis		Central Vietnam
R. jobiensis		Islands in Geelvink Bay, New Guinea; ref. 9.33
R. koratensis (*sladeni*) (*sikkimensis*)	Sladen's rat	Nepal – Indochina; forest
R. korinchi		Sumatra; montane; ref. 9.37
R. leucopus	Mottle-tailed rat (Cape York rat) (White-footed spiny rat)	S New Guinea, N Queensland; lowland forest
R. losea (*exiguus*) (*sakeratensis*)	Lesser ricefield rat	SE China – Thailand, Taiwan; cultivation
R. lugens		Mentawai Is; ref. 9.106
R. lutreolus	Australian swamp rat	E, SE Australia, Tasmania
R. macleari†		Christmas I, Indian Oc.; extinct
R. marmosurus (*facetus*)		Sulawesi
R. mindorensis		Mindoro, Philippines; ref. 9.106
R. montanus		Sri Lanka; montane
R. morataiensis		Moluccas
R. mordax		E New Guinea; ref. 9.33
R. nativitatis†		Christmas I, Indian Ocean; extinct

R. niobe	Moss-forest rat	New Guinea; montane
R. nitidus (*ruber*)	Himalayan rat	Himalayas – Indochina, Luzon (Philippines), [Sulawesi, New Guinea]; commensal
R. norvegicus	Brown rat (Common rat) (Norway rat)	[Worldwide], original range probably SE Asia; commensal; includes domestic rats
R. novaeguinae		NE New Guinea; ref. 9.33
R. osgoodi		S Vietnam; ref. 9.84
R. owiensis		Owi I, W New Guinea
R. palmarum		Nicobar Is
R. praetor (*ruber*)	Variable spiny rat	New Guinea, Solomon Is
R. ranjiniae		S India
R. rattus	Roof rat (Black rat) (Ship rat) (House rat)	SE Asia, [worldwide in tropics and warm temperate zones]; commensal
R. richardsoni (*omichlodes*)	Richardson's rat	W New Guinea; montane
R. sordidus (*colletti*)	Australian dusky field rat (Canefield rat)	N, NE Australia, New Guinea
R. steini (*leucopus*)		New Guinea; ref. 9.33
R. stoicus (*rogersi*)		Andaman Is
R. tawitawiensis		Tawitawi I, S Philippines; ref. 9.106
R. tiomanicus (*jalorensis*)	Malaysian field rat	Malaya – Borneo, Palawan; cultivation, secondary forest
R. tunneyi	Tunney's rat (Pale field rat)	W, N, NE, C Australia; savanna, coastal sands
R. turkestanicus (*rattoides*)	Turkestan rat	Soviet Turkestan — China
R. verecundus	Slender rat	New Guinea
R. villosissimus	Long-haired rat	E, C Australia
R. xanthurus		Sulawesi

Maxomys; (*Rattus*); oriental spiny rats; SE Asia; ref. 9.30.

M. alticola	Mountain spiny rat	N Borneo; montane forest
M. baeodon	Small spiny rat	N Borneo
M. bartelsii		Java
M. dollmani		Sulawesi
M. hellwaldii		Sulawesi
M. hylomyoides		Sumatra
M. inas	Malayan mountain spiny rat	Malaya; montane forest

M. inflatus		Sumatra
M. moi		Indochina
M. musschenbroekii		Sulawesi
M. ochraceiventer	Chestnut-bellied spiny rat	Borneo; submontane forest
M. pagensis		Mentawai Is
M. panglima		Palawan etc.
M. rajah	Brown spiny rat (Rajah rat)	Malaya – Sumatra, Borneo; lowland forest
M. surifer	Red spiny rat	Indochina – Java, Borneo; lowland forest
M. whiteheadi	Whitehead's rat	Malaya, Sumatra, Borneo; forest

Sundamys; (*Rattus*); ref. 9.34.

S. infraluteus	Sumatra, Borneo
S. maxi	Java
S. muelleri	Malaya, Sumatra, Borneo, Palawan

Abditomys; (*Rattus*); ref. 9.35.

A. latidens	Luzon, Philippines

Palawanomys; ref. 9.34.

P. furvus	Palawan, Philippines

Niviventer; (*Rattus, Maxomys*); SE Asia; ref. 9.31.

N. andersoni		SW China
N. brahma		NE India, N Burma; montane forest
N. bukit	Malayan white-bellied rat	S Burma – Java; forest
N. confucianus	Chinese white-bellied rat	S Manchuria – N Indochina, Taiwan, Hainan
N. coninga (*coxingi*)		N Burma, Taiwan
N. cremoriventer	Dark-tailed tree rat	Malaya – Java, Borneo; lowland forest
N. eha	Smoke-bellied rat	Nepal – SW China; montane
N. excelsior		SW China
N. fulvescens (*bukit*)	Chestnut rat	Himalayas – S China – Java; ref. 9.29
N. hinpoon		Thailand
N. langbianus		NE India – Thailand, Indochina; close to *N. cremoriventer*
N. lepturus		Java
N. niviventer	White-bellied rat	Himalayas
N. rapit	Long-tailed mountain rat	Thailand – Sumatra, Borneo
N. tenaster		Assam – Vietnam

Margaretamys; (*Rattus*); Sulawesi; forest; ref. 9.31.

M. beccarii		Sulawesi; lowland
M. elegans		Sulawesi; montane
M. parvus		Sulawesi; montane

Berylmys; (*Rattus*); ref. 9.34.

B. berdmorei	Small white-toothed rat (Grey rat)	S Burma, Thailand, Indochina
B. bowersi	Bower's rat	NW India – S China – Sumatra; forest
B. mackenziei	Kenneth's white-toothed rat	Assam – Thailand
B. manipulus	Manipur rat	Assam – Burma

Lenothrix; (*Rattus*).

L. canus	Grey tree rat	Malaya, Sumatra, Borneo; forest

Apomys; (*Rattus*); Philippines; ref. 9.112.

A. abrae	Luzon
A. datae	Luzon
A. hollisteri (*microdon*)	Leyte, Dinagat, etc.
A. hylocoetes (*petraeus*)	Mindanao
A. insignis	Mindanao
A. littoralis	Mindanao, Negros
A. musculus	Luzon, Mindoro
A. sacobianus	Luzon

Diplothrix; (*Rattus*).

D. legatus	Ryukyu Is

Leopoldamys; (*Rattus*); ref. 9.31.

L. edwardsi	Edwards' rat	Sikkim – S China – Sumatra
L. neilli	Neill's rat	S Thailand
L. sabanus	Long-tailed giant rat (Noisy rat)	Indochina – Java, Borneo; lowland forest
L. siporanus		Mentawai Is

Bunomys; (*Rattus*); ref. 9.32.

B. andrewsi (*adspersus*)	Sulawesi
B. chrysocomus	Sulawesi
B. fratrorum	Sulawesi
B. penitus	Sulawesi

Taeromys; (*Rattus*); ref. 9.32.

T. arcuatus	Sulawesi

T. callitrichus		Sulawesi
T. celebensis		Sulawesi
T. hamatus		C Sulawesi
T. punicans		Sulawesi
T. taerae		NE Sulawesi

Cremnomys; (*Rattus*); India, etc.

C. blanfordi		India, Sri Lanka
C. cutchicus		S, W India
C. elvira		SE India

Paruromys; (*Rattus*); ref. 9.32.

P. dominator		Sulawesi

Bullimus; (*Rattus*); Philippines; ref. 9.32.

B. bagobus (*bagopus*)		Mindanao etc.
B. luzonicus		Luzon
B. rabori		Mindanao

Srilankamys; (*Rattus*); ref. 9.31.

S. ohiensis	Ohiya rat	Sri Lanka; montane

Kadarsanomys; (*Rattus*); ref. 9.36.

K. sodyi		W Java; montane forest

Anonymomys; ref. 9.31.

A. mindorensis		Mindoro, Philippines

Aethomys; (*Rattus*); Africa S of Sahara; savanna.

A. bocagei		Angola, Zaire
A. chrysophilus	Red veld rat	SW Africa – Natal – Kenya
A. granti	Grant's rock mouse	Karoo, S Africa
A. hindei		N Nigeria – NE Tanzania
A. kaiseri		E Zaire – Kenya – Malawi – Angola
A. namaquensis	Namaqua rock mouse	S Africa – S Zambia
A. nyikae (*dollmani*)	Nyika veld rat	NE Zambia, Malawi, E Zimbabwe
A. silindensis	Silinda rat	E Zimbabwe
A. thomasi		W C Angola

Thallomys; (*Aethomys, Rattus*).

T. paedulcus	Acacia rat (Black-tailed tree rat)	S Africa – Somalia; savanna, dry forest

Praomys; (*Myomys*); African soft-furred rats; Africa (Arabia); mainly forest.

P. albipes		Ethiopia, etc.; forest, cultivation

P. daltoni (*butleri*)		Senegal – Sudan; savanna
P. delectorum		Kenya – Malawi; montane forest
P. derooi		Ghana – W Nigeria; savanna; ref. 9.38
P. fumatus		E Africa, SW Arabia; steppe
P. hartwigi		Cameroun; montane forest
P. jacksoni		Nigeria – Sudan – Zambia; forest
P. misonnei		E Zaire; ref. 9.107
P. morio		W C, Africa; forest
P. rostratus		Ivory Coast, Liberia; ref. 9.40
P. ruppi		Ethiopia; montane; ref. 9.41
P. tullbergi		Senegal – Ghana; forest

Mastomys; (*Praomys, Rattus*); multimammate mice; Africa.

M. angolensis		Angola, Zaire

M. coucha
  (*natalensis*)
		S, ? E, ? W Africa; close to
		  M. natalensis

M. erythroleucus
  (*natalensis*)
		Morocco, Senegal – Sudan;
		  steppe, savanna,
		  cultivation

M. huberti
  (*erythroleucus*)
		W Africa, Morocco,
		  ? Somalia; guinea
		  savanna; ref. 9.39

M. natalensis	Natal multimammate rat	S, ? E Africa; steppe,
		  savanna, cultivation,
		  commensal; close to *M.*
		  coucha

M. pernanus		S Kenya, N Tanzania, Ruanda
M. shortridgei	Shortridge's mouse	Namibia, Botswana
M. verheyeni		Nigeria, Cameroun; savanna; ref. 9.108

Myomyscus; (*Praomys*).

M. verroxii (*verreauxi*)	Verreaux's mouse	Cape Province

Hylomyscus; (*Praomys, Rattus*); Africa S of Sahara; forest; classification very provisional.

H. alleni	Guinea – Gabon, etc.
H. baeri	Ivory Coast, Ghana
H. carillus (*aeta*)	Zaire, etc.

H. denniae		Uganda – Zambia
H. fumosus		Gabon, Cameroun, etc.
H. parvus		Zaire – Cameroun
H. stella		Nigeria – Kenya

Limnomys; (*Rattus*).

| *L. sibuanus* | | Mindanao, Philippines; |
| (*mearnsi*) | | montane |

Stochomys; (*Rattus*); ref. 9.42.

| *S. longicaudatus* | | Zaire – Nigeria; rain forest |

Dephomys; (*Stochomys*); tropical Africa; forest; ref. 9.42.

D. defua		Guinea – Ivory Coast
D. eburnea		Ivory Coast
(*defua*)		

Tarsomys; (*Rattus*).

| *T. apoensis* | | Mindanao, Philippines |

Tryphomys; (*Rattus*).

| *T. adustus* | Mearns's Luzon rat | Luzon, Philippines; |
| | | montane |

Leporillus

L. apicalis	White-tipped stick-nest rat	C Australia; dry forest,
	(Lesser stick-nest rat)	scrub; probably extinct
L. conditor	Greater stick-nest rat	Franklin I, S Australia
		(extinct mainland); R

Leggadina; (*Pseudomys*); ref. 9.43.

| *L. forresti* | Forrest's mouse | W, C Australia; arid zones |
| *L. lakedownensis* | Lakeland Downs mouse | NE Queensland |

Pseudomys; (*Gyomys, Thetamys*); Australian mice; Australia; forest – desert.

P. albocinereus	Ashy-grey mouse	SW Australia; scrub
P. apodemoides	Silky mouse	S Australia, Victoria
(*albocinereus*)		
P. australis	Eastern mouse	E, S, C Australia
	(Plains rat)	
P. bolami		SW Australia; ref. 9.110
(*hermanns-*		
burgensis)		
P. chapmani	Pebble-mound mouse	NW Australia; ref. 9.44
P. delicatulus	Little native mouse	NE, N, NW Australia,
	(Delicate mouse)	C New Guinea; steppe
P. desertor	Brown desert mouse	W, C Australia
P. fieldi	Alice Springs mouse	Alice Springs, C Australia;
		? extinct
P. fumeus	Smokey mouse	Victoria; forest

P. glaucus (*albocinereus*)		S Queensland
P. gouldii	Gould's mouse	SW, S Australia; ? extinct
P. gracilicaudatus	Eastern chestnut mouse	C, S Queensland, New South Wales
P. hermanns-burgensis	Sandy inland mouse	Australia; dry grassland, steppe
P. higginsi	Tasmanian mouse (Long-tailed mouse)	Tasmania; forest
P. johnsoni		N Australia; ref. 9.109
P. laborifex		NW Australia; ref. 9.111
P. nanus	Western chestnut mouse	W, N Australia; rocky hills
P. novaehollandiae	New Holland mouse	E New South Wales, Victoria, Tasmania
P. occidentalis	Western mouse	SW Australia; scrub
P. oralis	Hastings River mouse	SE Queensland; I
P. pilligaensis	Pilliga mouse	N New South Wales; ref. 9.45
P. praeconis	Shark Bay mouse	Shark Bay, W Australia; coastal dunes; R
P. shortridgei	Heath rat	Victoria, ? SW Australia; swamp, scrub

Melomys; (*Uromys*); mosaic-tailed rats; New Guinea, NE Australia; mainly montane forest.

M. aerosus		Ceram; montane
M. albidens	White-toothed melomys	W New Guinea; montane
M. arcium	Rossel Island melomys	Rossel I, E of New Guinea
M. burtoni (*littoralis*)	Grassland melomys	E, N Australia
M. capensis (*cervinipes*)	Cape York melomys	NE Queensland
M. caurinus		Talaud Is, Moluccas
M. cervinipes	Fawn-footed melomys	E, NE Australia; forest
M. fellowsi	Red-bellied melomys	NE New Guinea; montane
M. fraterculus		Ceram; montane
M. fulgens		Ceram
M. hadrourus	Thornton Peak melomys	NE Queensland; ref. 9.46
M. leucogaster	White-bellied melomys	New Guinea
M. levipes	Long-nosed melomys	New Guinea
M. lorentzii	Long-footed melomys	New Guinea
M. lutillus	Little melomys	SE New Guinea, N Queensland
M. monktoni	Southern melomys	New Guinea
M. obiensis		Obi I, Halmahera
M. platyops	Lowland melomys	New Guinea
M. porculus		Guadalcanal, Solomon Is
M. rubex	Highland melomys	New Guinea

| *M. rubicola* | Bramble Cay melomys | Bramble Cay (I), E New Guinea; (in *M. leucogaster?*) |
| *M. rufescens* | Rufescent melomys | New Guinea – Solomon Is |

Pogonomelomys; (*Melomys*); New Guinea; forest; arboreal, terrestrial.

P. bruijnii	Lowland brush mouse	W, S New Guinea; lowland
P. mayeri	Shaw Mayer's brush mouse	W New Guinea; montane
P. ruemmleri	Rümmler's brush mouse	W New Guinea; montane
P. sevia	Highland brush mouse	NE New Guinea; montane

Solomys; (*Melomys*); naked-tailed rats; Solomon Islands; forest.

S. ponceleti		Bougainville I, Solomons
S. salebrosus		Bougainville I, Solomons
S. sapientis		Santa Ysabel I, Solomons

Uromys; giant naked-tailed rats; New Guinea, Solomon Islands, NE Australia; forest; arboreal.

U. anak (*neobritannicus*)	Black-tailed tree rat	C, NE New Guinea; New Britain; montane
U. caudimaculatus	Giant white-tailed rat (Mottle-tailed tree rat)	New Guinea, NE Queensland
U. imperator		Guadalcanal I, Solomons
U. rex		Guadalcanal I, Solomons
U. salamonis		Florida I, Solomons

Giant white-tailed rat (Uromys caudimaculatus)

Xenuromys

| *X. barbatus* | Mimic tree rat | New Guinea |

Malacomys; African swamp rats; W, C Africa; forest, swamp.

M. cansdalei		Ivory Coast – Ghana
M. edwardsi		Liberia – Nigeria
M. longipes		Cameroun – Uganda – NW Zambia
M. verschureni		E, NW Zaire

Haeromys; pygmy tree mice; Borneo, Sulawesi; forest.

H. margarettae	Ranee mouse	Borneo
H. minahassae		N Sulawesi
H. pusillus	Lesser ranee mouse	Borneo

Chiromyscus

| *C. chiropus* | Fea's tree rat | Indochina, E Burma; forest |

Diomys

| *D. crumpi* | Crump's mouse | NE India, Nepal |

Zelotomys

| *Z. hildegardeae* | | Angola – Malawi – Uganda – C Af. Rep. |

Z. woosnami	Pale rat (Woosnam's desert rat)	NW Botswana, EC Namibia

Muriculus; (*Mus*).
M. imberbis   Ethiopia; montane

Mus; (*Coelomys, Gatimiya, Leggada, Leggadilla, Mycteromys, Nannomys*); SE Asia, Africa, [worldwide]; forest, grassland; ref. 9.49 (Asia); 9.48 (Europe); 9.50.

House mouse (Mus musculus)

M. abbotti	Abbott's mouse	Asia Minor, Greece, etc.
M. baoulei		Ivory Coast; ref. 9.47
M. booduga (*fulvidiventris*) (*lepidoides*)		India, Pakistan, Burma, Sri Lanka
M. bufo		E Zaire
M. callewaerti		S Zaire, Angola
M. caroli	Ryukyu mouse	Ryukyu Is, Taiwan, S China, Indochina – Sumatra, Java; cultivation
M. cervicolor	Fawn-coloured mouse	Kashmir – Indochina, Sumatra, Java; cultivation, forest
M. cookii	Cook's mouse	India – Thailand
M. crociduroides		Sumatra; montane forest
M. dunni		India, Sumatra
M. famulus		S India; montane forest
M. fernandoni		Sri Lanka
M. goundae		N Cent. African Rep.
M. gratus		W Uganda
M. haussa		Senegal – Niger; steppe
M. hortulanus (*musculus*)	Steppe mouse	Austria – S Russia; cultivated land
M. indutus	Desert pygmy mouse	S Africa – Zambia
M. mahomet		Ethiopia
M. mattheyi		Senegal – Ghana; savanna
M. mayori		Sri Lanka
M. minutoides (*musculoides*)	Pygmy mouse	Africa S of Sahara; probably an aggregate of sibling species
M. musculus (*molossinus*) (*domesticus*) (*castaneus*) (*poschiavinus*)	House mouse	worldwide commensal, outdoor forms scarce in range of other *Mus* spp.
M. oubanguii		Cent. African Rep.; savanna
M. pahari	Gairdner's shrew-mouse	Sikkim – Indochina
M. phillipsi		India
M. platythrix		India

M. proconodon (*pasha*)		Somalia, Ethiopia, Zaire
M. saxicola		India, Pakistan, Nepal
M. setulosus		Guinea – Gabon; Ethiopia
M. setzeri	Setzer's pygmy mouse	Botswana, Zambia
M. shortridgei	Shortridge's mouse	Burma – Indochina
M. sorella (*neavei*)		Uganda, Kenya – Zambia
M. spicilegus	Steppe mouse	SE Europe – Iran
M. spretus	Algerian mouse	SW Europe, NW Africa
M. tenellus		E Africa
M. triton	Grey-bellied pygmy mouse	E Zaire – Kenya – Mozambique
M. vulcani		Java; montane forest

Colomys; (*Nilopegamys*).

C. goslingi		Angola – Cameroun – Ethiopia; wet forest

Crunomys; ref. 9.112.

C. celebensis		C Sulawesi
C. fallax		Luzon, Philippines
C. melanius		Mindanao, Philippines
C. rabori		Leyte, Philippines

Archboldomys; ref. 9.112.

A. luzonensis		Luzon, Philippines

Macruromys

M. elegans	Western small-toothed rat	W New Guinea; montane
M. major	Eastern small-toothed rat	NE New Guinea; montane

Lorentzimys

L. nouhuysii	New Guinea jumping mouse	New Guinea; forest

Lophuromys; brush-furred rats; W, C, E Africa; forest, etc.

L. cinereus		E Zaire; montane
L. flavopunctatus		C, E Africa, Ethiopia
L. luteogaster		NE Zaire; forest
L. medicaudatus		C Africa; montane
L. melanonyx		Ethiopia; high montane
L. nudicaudus		Cameroun, etc.; forest-edge
L. rahmi		C Africa; montane forest
L. sikapusi		W, C Africa; forest savanna
L. woosnami		C Africa; montane forest

Notomys; Australian hopping mice; Australia; desert, steppe, (dry forest).

N. alexis	Spinifex hopping mouse	C, W Australia; desert, dry grassland
N. amplus	Short-tailed hopping mouse	C Australia; probably extinct
N. aquilo (*carpentarius*)	Northern hopping mouse	N Queensland, N Australia; coastal dunes; K
N. cervinus	Fawn hopping mouse	C Australia; desert
N. fuscus	Dusky hopping mouse	C Australia; desert, steppe; I
N. longicaudatus	Long-tailed hopping mouse	SW, C Australia; ? extinct
N. macrotis (*megalotis*)	Big-eared hopping mouse	SW Australia; ? extinct
N. mitchellii	Mitchell's hopping mouse	SE, S, SW Australia; dry forest – grassland

Mastacomys

M. fuscus	Broad-toothed rat	SE Australia, Tasmania; wet forest, marsh

Echiothrix

E. leucura	Sulawesi spiny rat (Celebes shrew-rat)	Sulawesi

Melasmothrix

M. naso	Lesser shrew-rat	Sulawesi

Tateomys; shrew rats; Sulawesi; ref. 9.51.

T. macrocercus		Sulawesi
T. rhinogradoides	Tate's rat	Sulawesi

Acomys; African spiny mice; Africa, SW Asia; desert – savanna; classification very provisional.

Cairo spiny mouse (Acomys cahirinus)

A. cahirinus (*cineraceus*) (*dimidiatus*)	Cairo spiny mouse	circum-Sahara, Ethiopia, Kenya, Israel – Pakistan; perhaps an aggregate of sibling species
A. cilicicus		Asia Minor; ref. 9.52
A. louisae (*subspinosus*)		Somalia; ref. 9.54
A. minous (*cahirinus*)		Crete
A. russatus	Golden spiny mouse	NE Egypt – Jordan, E Arabia
A. spinosissimus		Mozambique – Botswana
A. subspinosus		S Africa – Somalia, Sudan
A. whitei		Oman; ref. 9.53
A. wilsoni		Kenya, S Sudan, S Ethiopia

Uranomys
U. ruddi   Rudd's mouse   Senegal – Kenya –
                                                Mozambique; savanna

Bandicota; (*Gunomys*); bandicoot-rats; SE Asia; scrub, cultivation, partly
commensal.
B. bengalensis   Lesser bandicoot-rat   India – Burma, Sumatra,
                                                Java, Sri Lanka
B. indica   Greater bandicoot-rat   India – S China – Java,
                                                Sri Lanka, Taiwan
B. savilei   Burma – Indochina

Nesokia
N. indica   Short-tailed bandicoot-rat   Egypt – NW India,
                                                Sinkiang; steppe,
                                                cultivation

Erythronesokia; ref. 9.55.
E. bunni   S Iraq; marshes

Anisomys
A. imitator   New Guinea giant rat   New Guinea; montane
                          (Squirrel-toothed rat)   forest

Lenomys
L. meyeri   Sulawesi giant rat   Sulawesi
  (*longicaudus*)

Pogonomys; prehensile-tailed mice; New Guinea, etc.; forest; arboreal; ref. 9.56.
P. championi   C New Guinea; montane;
                                                ref. 9.126
P. loriae   Soft-haired tree mouse New Guinea,
  (*fergussoniensis*)   (Prehensile-tailed rat)   d'Entrecasteaux Is,
  (*mollipilosus*)   N Queensland; montane
P. macrourus   Long-tailed tree mouse New Guinea; lowland
P. sylvestris   Grey-bellied tree mouse New Guinea; montane

Chiruromys; (*Pogonomys*); New Guinea etc.; forest; arboreal; ref. 9.56.
C. forbesi   Greater tree mouse   E New Guinea, etc.
  (*shawmayeri*)   (Forbes' tree mouse)
C. lamia   Broad-skulled tree mouse   SE New Guinea
  (*kagi*)
C. vates   Lesser tree mouse   SE New Guinea; lowland

Chiropodomys; pencil-tailed tree mice; SE Asia; forest.
C. calamianensis   Palawan, etc., Philippines
C. gliroides   NE India – Java, Borneo
  (*pusillus*)
C. jingdongensis   Yunnan, China; montane;
                                                ref. 9.57.

C. karlkoopmani		Siberut & N Pagai Is, Mentawai Is
C. major		Borneo
C. muroides		Borneo

Mallomys; New Guinea; ref. 9.125

M. aroaensis		E New Guinea
M. gunung		C New Guinea; high montane
M. istapantap		C New Guinea
M. rothschildi	Smooth-tailed giant rat (Black-eared giant rat)	New Guinea; montane

Papagomys

P. armandvillei	Flores giant rat	Flores (Lesser Sunda Is)

Komodomys

K. rintjanus	Komodo rat	Komodo Is, Indonesia; ref. 9.58

Phloeomys; slender-tailed cloud rats; Luzon, Philippines; forest; ref. 9.112.

P. cumingi		Luzon, Philippines
P. pallidus		Luzon, Philippines; (in *P. cumingi* ?)

Crateromys; bushy-tailed cloud rats; Philippines; forest; ref. 9.59.

C. australis		Dinagat I (Philippines); ref. 9.112
C. paulus		Ilin I, N Philippines; ? extinct
C. schadenbergi		Luzon, Philippines; montane forest

Chrotomys; Philippine striped rats; Philippines; ref. 9.60.

C. mindorensis	Mindoro striped rat	Mindoro, Luzon; lowland
C. whiteheadi	Luzon striped rat	Luzon, Philippines; montane

Celaenomys

C. silaceus	Luzon shrew-rat	Luzon, Philippines; montane forest

Crossomys

C. moncktoni	Earless water rat	E New Guinea; streams

Xeromys

X. myoides	False swamp rat	N, NE Australia; swamps; K.

Hydromys; (*Baiyankamys*); beaver-rats; Australia, New Guinea, etc.; rivers; predators.

H. chrysogaster	Beaver-rat (Australian water rat)	Australia, Tasmania, New Guinea
H. habbema	Mountain water rat	W New Guinea
H. hussoni		W, C New Guinea; ref. 9.61
H. neobritannicus		New Britain

Parahydromys

P. asper	Coarse-haired hydromyine	New Guinea; montane forest

Neohydromys

N. fuscus	Short-tailed shrew-mouse	NE New Guinea; high montane

Leptomys

L. elegans	Long-footed hydromyine	New Guinea

Paraleptomys; New Guinea; montane forest.

P. rufilatus	Red-sided hydromyine	N New Guinea
P. wilhelmina	Short-footed hydromyine (Short-haired hydromyine)	W New Guinea

Pseudohydromys; New Guinea; high montane forest.

P. murinus	Eastern shrew-mouse	NE New Guinea
P. occidentalis	Western shrew-mouse	W New Guinea

Microhydromys

M. richardsoni	Groove-toothed shrew-mouse	W New Guinea

Mayermys

M. ellermani	One-toothed shrew-mouse	NE New Guinea; montane forest

Rhynchomys; Luzon, Philippines; montane forest; ref. 9.62.

R. isarogensis	Isarog shrew-rat	Mt Isarog, Luzon
R. soricoides	Mount Data shrew-rat	Luzon

Family Gliridae (Muscardinidae)

Dormice; *c.* 20 species; Palaearctic, Africa; mainly forest, eating seeds and buds.

Glis; (*Myoxus*).

G. glis	Fat dormouse (Edible dormouse)	Europe, Asia Minor; forest

Muscardinus

M. avellanarius	Hazel dormouse	Europe, Asia Minor

Beaver-rat
(Hydromys chrysogaster)

Japanese dormouse
(Glirurus japonicus)

Eliomys

E. melanurus		N Africa, Asia Minor – N Arabia
E. quercinus	Garden dormouse	Europe

Dryomys

D. laniger	Woolly dormouse	SW Asia Minor
D. nitedula	Forest dormouse	E Europe – Sinkiang

Glirurus

G. japonicus	Japanese dormouse	Japan, except Hokkaido

Chaetocauda

C. sichuanensis	Chinese dormouse	N Sichuan, China; ref. 9.82

Myomimus; mouse-tailed dormice.

M. personatus		N Iran, etc.
M. roachi		Bulgaria, etc., ?† Asia Minor, etc.
M. setzeri		Iran

Graphiurus; (*Claviglis*); African dormice; Africa S of Sahara; forest, savanna; classification very provisional; ref. 9.121.

G. christyi		NE Zaire – Cameroun
G. crassicaudatus		Liberia – Cameroun
G. hueti		Guinea – Angola
G. lorraineus		Senegal – E Zaire
G. monardi		NW Zambia etc.
G. murinus	Woodland dormouse	Africa S of Sahara
G. ocularis	Spectacled dormouse	SW Africa
G. parvus	Lesser savanna dormouse	Sierra Leone – Somalia – Zimbabwe
G. platyops	Rock dormouse	Zimbabwe – S Zaire – Namibia

Family Seleviniidae

One species.

Desert dormouse (Selevinia betpakdalaensis)

Selevinia

S. betpakdalaensis	Desert dormouse	SE Kazakhstan; desert

Family Zapodidae

Jumping mice; *c.* 17 species; N Eurasia, N America; forest, grassland.

Sicista; birch mice; N Eurasia; forest, grassland.

S. armenica		Caucasus; ref. 9.63
S. betulina	Northern birch mouse	N, C Europe – E Siberia; forest
S. caucasica		Caucasus

Northern birch mouse (Sicista betulina)

S. concolor	Chinese birch mouse	Altai, Kashmir, W China; montane
S. kasbegica		Caucasus; ref. 9.113
S. kluchorica		W Caucasus; ref. 9.113
S. napaea	Altai birch mouse	N Altai Mts
S. pseudonapaea		S Altai Mts
S. severtzovi (*subtilis*)		S Russia; ref. 9.113
S. strandi		S Russia, Caucasus; ref. 9.123
S. subtilis	Southern birch mouse	E Europe – L Baikal; grassland
S. tianshanica	Tien Shan birch mouse	Tien Shan Mts

Zapus; American jumping mice; N America; wet grassland, forest-edge.

Z. hudsonius	Meadow jumping mouse	Canada, N, C USA
Z. princeps	Western jumping mouse	W North America
Z. trinotatus	Pacific jumping mouse	W coast USA; I

Eozapus

E. setchuanus	Sichuan jumping mouse	W, SW China; montane forest

Napaeozapus

N. insignis	Woodland jumping mouse	SE Canada, NE USA; forest

Family Dipodidae

Jerboas; *c.* 30 species; C Asia – NW Africa; desert, steppe.

Dipus

D. sagitta	Northern three-toed jerboa	S European Russia – N Iran – China

Lesser Egyptian jerboa (Jaculus jaculus)

Paradipus

P. ctenodactylus	Comb-toed jerboa	SW Turkestan

Jaculus

J. blanfordi (*turcmenicus*)	Blanford's jerboa	Turkmenia – W Pakistan
J. jaculus (*deserti*)	Lesser Egyptian jerboa	Sahara, Arabia; desert
J. lichtensteini	Lichtenstein's jerboa	Caspian Sea – L Balkhash
J. orientalis	Greater Egyptian jerboa	Morocco – Israel; semidesert

Stylodipus; Thick-tailed three-toed jerboas; ref. 9.114.

S. andrewsi		Mongolia
S. sungorus		W Mongolia
S. telum		Ukraine – Kazakhstan

Allactaga

A. bobrinskii	Bobrinski's jerboa	Kizil-kum, Kara-kum Deserts, Turkestan
A. bullata		S, W Mongolia, etc.
A. elater	Small five-toed jerboa	E Asia Minor – Baluchistan; dry steppe
A. euphratica (*williamsi*)	Euphrates jerboa	Jordan – Afghanistan; dry steppe
A. firouzi		Isfahan, Iran; ref. 9.80
A. hotsoni	Hotson's jerboa	Persian Baluchistan
A. major	Great jerboa	Ukraine – Tien Shan
A. nataliae		Mongolia; ref. 9.64
A. severtzovi	Severtzov's jerboa	Turkestan
A. sibirica	Mongolian five-toed jerboa	R Ural – Manchuria; steppe
A. tetradactyla	Four-toed jerba	Libya, Egypt; coastal gravel plains

Alactagulus

A. pumilio (*pygmaeus*)		R Don – Inner Mongolia, N Iran

Pygeretmus; fat-tailed jerboas.

P. platyurus (*vinogradovi*)	Lesser fat-tailed jerboa	Kazakhstan
P. shitkovi	Greater fat-tailed jerboa	E Kazakhstan

Cardiocranius

C. paradoxus	Five-toed pygmy jerboa	Mongolia, etc.

Salpingotus; (*Salpingotulus*); three-toed pygmy jerboas; Mongolia – Baluchistan; ref. 9.65.

S. crassicauda	Thick-tailed pygmy jerboa	S Mongolia, E Turkestan; desert
S. heptneri	Heptner's pygmy jerboa	Uzbekistan, S of Aral Sea
S. kozlovi	Koslov's pygmy jerboa	S Mongolia; desert
S. michaelis	Baluchistan pygmy jerboa	NW Baluchistan
S. pallidus		Kazakhstan

Euchoreutes

E. naso	Long-eared jerboa	W Sinkiang – Inner Mongolia; desert

Family Hystricidae

Old-world porcupines; *c.* 11 species; Africa, S Asia; forest, savanna, steppe; ground vegetarians; ref. 9.66.

Hystrix; (*Acanthion*, *Thecurus*); short-tailed porcupines; Africa, S Asia, (S Europe); forest, savanna.

H. africaeaustralis		S, SE Africa; savanna

*Crested porcupine
(Hystrix cristata)*

H. brachyura (*hodgsoni*)		Nepal – E China – Malaya, Sumatra, Borneo; forest
H. crassispinis		Borneo
H. cristata	Crested porcupine	W, E, N Africa, (S Europe); savanna, etc.
H. indica		SW Asia – India, Sri Lanka; steppe, scrub
H. javanica		Java – Flores, [Sulawesi ?]; forest
H. pumilis		Palawan, Philippines
H. sumatrae		Sumatra

Atherurus; brush-tailed porcupines; Africa, SE Asia; forest.

A. africanus	African brush-tailed porcupine	W, C, E Africa
A. macrourus	Asiatic brush-tailed porcupine	Assam – Malaya

Trichys

T. fasciculata (*lipura*)	Long-tailed porcupine	Malaya, Sumatra, Borneo; forest

Family Erethizontidae

New-world porcupines *c.* 10 species; N, C, S America; forest.

Erethizon

E. dorsatum	North American porcupine	Alaska – N Mexico; forest

Coendou; (*Sphiggurus*); tree porcupines; Mexico – Bolivia; forest.

C. bicolor (*rothschildi*)		Bolivia – Panama
C. insidiosus		E, C Brazil
C. mexicanus	Mexican porcupine	S Mexico – W Panama
C. prehensilis		N Argentina, Brazil – Venezuela
C. spinosus		E Brazil – N Argentina
C. vestitus (*pruinosus*)		Venezuela, Colombia
C. villosus		SE Brazil

Tree porcupine (C. prehensilis)

Echinoprocta

E. rufescens	Upper Amazonian porcupine	Colombia; forest

Chaetomys

C. subspinosus	Thin-spined porcupine	E, N Brazil; scrub; I

Guinea pig
(Cavia tschudii)

Family Caviidae

Guinea pigs, etc.; *c.* 16 species; S America.

Cavia; guinea pigs, cavies; Colombia – N Argentina; delimitation of species very provisional.

C. anolaimae		Colombia
C. aperea		N Argentina – E Brazil
C. fulgida		E Brazil
C. guianae		
C. magna		Surinam – Venezuela S Brazil, Uruguay; ref. 9.67
C. nana		W Bolivia
C. tschudii		Ecuador – N Argentina; probable ancestor of domestic Guinea pig, *C. porcellus*

Kerodon

K. rupestris	Rock cavy	NE Brazil

Galea

G. flavidens		Brazil
G. musteloides		Peru – C Argentina
G. spixii (*wellsi*)		Brazil, E Bolivia

Microcavia

M. australis		Argentina
M. niata		Bolivia; montane
M. shiptoni		NW Argentina; montane

Dolichotis; (*Pediolagus*); maras (Patagonian cavies).

D. patagonum		Argentina
D. salinicola		NE Argentina, Paraguay, S Bolivia

Family Hydrochaeridae

One species.

Hydrochaerus

H. hydrochaeris (*isthmius*)	Capybara	E Argentina – Panama

Capybara
(Hydrochaerus hydrohaeris)

Family Dinomyidae

One species.

Dinomys

D. branickii	Pacarana	Venezuela – Bolivia; forest; ref. 9.115

Pacarana
(Dinomys branickii)

Family Dasyproctidae

Pacas, agoutis; *c.* 14 species; Mexico – S Brazil; forest, savanna.

Agouti; (*Cuniculus, Coelogenys, Stictomys*).

A. paca		S Mexico – Surinam – Paraguay, [Cuba]
A. taczanowskii		Venezuela – Ecuador; montane

Agouti
(Agouti paca)

Dasyprocta; agoutis; C, S America; forest, savanna.

D. azarae		C Brazil – N Argentina
D. coibae		Coiba I, Panama
D. cristata		Surinam
D. fuliginosa	Grey agouti	Colombia, upper Amazon, Surinam
D. guamara		Venezuela; swamp forest
D. kalinowskii		Peru
D. leporina (*albida*) (*aguti*)	Brazilian agouti	Venezuela – E Brazil, Lesser Antilles
D. mexicana	Mexican agouti	Veracruz, S Mexico, [Cuba]
D. punctata (*variegata*)		S Mexico – N Argentina, [Cuba]
D. ruatanica		Roatin I, Honduras

Myoprocta; acuchis; Amazon Basin; forest.

M. acouchy (*pratti*)		Amazon Basin
M. exilis		Amazon Basin

Family Chinchillidae

Viscachas, chinchillas; *c.* 6 species; southern S America.

Lagostomus

L. maximus	Plains viscacha	Argentina; grassland

Chinchilla
(Chinchilla laniger)

Lagidium; mountain viscachas; Peru – Patagonia; montane.

L. peruanum		Peru
L. viscacia		Bolivia – N Chile, W Argentina
L. wolffsohni		S Chile, SW Argentina

Chinchilla; chinchillas.

C. brevicaudata		Peru – N Argentina; I
C. lanigera		Bolivia, Chile; montane rocks; I

Hispaniolan hutia
(Plagiodontia aedium)

Family Capromyidae

Hutias, etc.; 14 species; West Indies; forest, (freshwater margins).

Capromys; (*Geocapromys, Mesocapromys, Mysateles*); Cuban hutias; Cuba, Jamaica, etc.

C. angelcabrerae		Cay Ana Maria, S Cuba; ref. 9.68; E
C. arboricolus		Cuba; arboreal; ref. 9.69
C. auritus		Cay Fragoso, N Cuba; E
C. brownii	Brown's hutia	Jamaica, Swan I; I
C. garridoi		Canarreos Arch., Cuba; I
C. gundlachi		I de Juventud, Cuba; ref. 9.116
C. ingrahami	Bahaman hutia	Plana Keys, Bahamas; R
C. melanurus	Bushy-tailed hutia	E Cuba; arboreal; I
C. meridionalis		I de Juventud, Cuba; ref. 9.116
C. nanus	Dwarf hutia	Cuba; E
C. pilorides	Desmarest's hutia	Cuba
C. prehensilis	Prehensile-tailed hutia	Cuba; arboreal
C. sanfelipensis		Cay Juan Garcia, etc., W Cuba; E

Plagiodontia

P. aedium	Hispaniolan hutia	Hispaniola; I
(*hylaeum*)		

Family Myocastoridae

Coypu
(Myocastor coypus)

Myocastor

M. coypus	Coypu (Nutria)	Chile, Argentina – Bolivia, S Brazil; [Europe, N Asia, E Africa]

Family Octodontidae

Degus, etc.; *c.* 9 species; southern S America; steppe, mainly montane.

Octodon; degus; Peru, Chile.

O. bridgesi		Chile
O. degus		W Peru, Chile
O. lunatus		Chile; coastal hills

Degu
(Octodon degus)

Octodontomys

O. gliroides	Mountain degu	Bolivia, N Chile, NW Argentina; montane steppe

Spalacopus

S. cyanus	Coruro	Chile; montane
(*tabanus*)		

Aconaemys
A. fuscus	Chilean rock rat	SC Chile, W Argentina;
(porteri)		forest; subterranean
A. sagei		Patagonia; ref. 9.18

Octomys; (*Tympanoctomys*).
O. barrerae		C Argentina
O. mimax	Viscacha-rat	W Argentina; montane

Family Ctenomyidae

Tuco-tucos; *c.* 38 species; Peru – Patagonia; subterranean; taxonomy very provisional.

Tuco-tuco
(Ctenomys)

Ctenomys
C. argentinus	Chaco, Argentina; ref. 9.71
C. australis	E Argentina; coastal
C. azarae	C Argentina
C. boliviensis	Bolivia, N Argentina
C. bonettoi	Chaco, Argentina; ref. 9.70
C. brasiliensis	E Brazil
C. colburni	SW Argentina
C. conoveri	Paraguay, N Argentina, Bolivia
C. dorsalis	Paraguay
C. emilianus	WC Argentina
C. flamarioni	S Brazil; ref. 9.72
C. frater	NW Argentina, S Bolivia
C. fulvus	W Argentina, N Chile
(robustus)	
C. haigi	Argentina; ref. 9.117
C. knighti	W Argentina
C. latro	N Argentina
C. leucodon	Bolivia, Peru
C. lewisi	S Bolivia
C. magellanicus	Patagonia, Tierra del Fuego
C. maulinus	Chile, Argentina
C. mendocinus	C Argentina
(haigi)	
C. minutus	S Brazil, Uruguay, N Argentina, Bolivia
C. nattereri	C Brazil
C. occultus	N Argentina
C. opimus	N Argentina, S Bolivia
C. perrensis	NE Argentina
C. peruanus	S Peru
C. pontifex	W Argentina
C. porteousi	C, E Argentina
C. saltarius	N Argentina
C. sericeus	SW Argentina

C. sociabilis	WC Argentina; ref. 9.117
C. steinbachi	E Bolivia
C. talarum	E Argentina
C. torquatus	Uruguay, NE Argentina
C. tuconax	NW Argentina
C. tucumanus	N Argentina
C. validus	C Argentina; ref. 9.73

Family Abrocomidae

Chinchilla-rats; 2 species; Peru – N Chile; grassland, scrub.

*Chinchilla rat
(Abrocoma bennetti)*

Abrocoma

A. bennetti	Chile
A. cinerea	S Peru – N Chile, NW Argentina

Family Echimyidae

American spiny rats; *c.* 45 species; C, S America; forest; taxonomy very provisional; ref. 9.74.

*Spiny rat
(Proechimys setosus)*

Proechimys; terrestrial spiny rats; C, S America; mainly forest.

P. albispinus		E Brazil
P. amphichoricus		S Venezuela, etc.
P. bolivianus		Upper Amazon
P. brevicauda		E Peru, NW Brazil
P. canicollis		Colombia – Venezuela
P. chrysaeolus		E Colombia
P. cuvieri		Guianas – Peru
P. decumanus		NW Peru, SW Ecuador; Pacific lowlands
P. dimidiatus		SE Brazil
P. goeldii		Amazonian Brazil
P. guairae		N Venezuela
P. gularis		Upper Amazon
P. guyannensis (*warreni*)		Brazil – E Colombia, Guianas
P. hoplomyoides		SE Venezuela, etc.
P. iheringi		E Brazil
P. longicaudatus		SW Brazil, etc.
P. magdalenae		Colombia
P. mincae		N Colombia
P. myosuros		E Brazil
P. oconnelli		E Colombia
P. oris		C Brazil
P. poliopus		NW Venezuela, etc.
P. quadruplicatus		Ecuador
P. semispinosus	Tomes' spiny rat	Honduras – Ecuador; Pacific coast only
P. setosus		E Brazil

P. simonsi		Peru – Colombia
(*hendeei*)		
P. steerei		Peru
P. trinitatis		Trinidad
P. urichi		N Venezuela

Hoplomys; doubtfully distinct from *Proechimys*.

H. gymnurus	Armoured rat	Honduras – Ecuador

Euryzygomatomys

E. spinosus	Guira	E, SE Brazil – N Argentina

Clyomys

C. bishopi	SE Brazil; ref. 9.76
C. laticeps	E, C Brazil, Paraguay

Carterodon

C. sulcidens	E Brazil

Thrichomys; (*Cercomys*).

T. apereoides	NE Brazil – Paraguay,
(*cunicularis*)	Bolivia

Mesomys

M. didelphoides	? Brazil
M. hispidus	Amazon Basin
M. obscurus	? Brazil
M. stimulax	N Brazil, Surinam

Lonchothrix

L. emiliae	Brazil S of Amazon

Isothrix; toros; N, C South America; river banks in forest; ref. 9.118.

I. bistriatus	Amazon & Orinoco basins
(*villosus*)	
I. pagurus	C Amazon
I. pictus	E Brazil

Diplomys; Panama – Ecuador; forest.

D. caniceps		Colombia, Ecuador
D. labilis	Gliding spiny rat	Panama
(*darlingi*)		
D. rufodorsalis		NE Colombia

Echimys; (*Makalata*); arboreal spiny rats; S America; forest.

E. armatus	Ecuador – Guianas, Trinidad
E. blainvillei	SE Brazil
E. braziliensis	E Brazil
E. chrysurus	NE Brazil, Surinam

E. dasythrix	SE, E Brazil
E. grandis	Amazon Basin
E. macrurus	Brazil S of Amazon
E. nigrispinus	E Brazil
E. saturnus	E Ecuador
E. semivillosus	Venezuela, Colombia
E. unicolor	? Brazil

Dactylomys; (*Lachnomys*); coro-coros; NW South America; forest; arboreal.

D. boliviensis	Bolivia, SE Peru
D. dactylinus	Colombia, Ecuador, Amazonian Brazil
D. peruanus	Peru; montane

Kannabateomys

K. amblyonyx	E Brazil – N Argentina

Olallamys; (*Thrinacodus*); ref. 9.119.

O. albicauda	Colombia
O. edax	W Venezuela

Family Thryonomyidae

Cane rats; *c.* 2 species; Africa S of Sahara; mainly marshes.

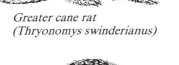

*Greater cane rat
(Thryonomys swinderianus)*

Thryonomys

T. gregorianus	Lesser cane rat	Cameroun – S Sudan – Zimbabwe
T. swinderianus	Greater cane rat	Africa S of Sahara

Family Petromuridae

One species.

*Dassie-rat
(Petromus typicus)*

Petromus

P. typicus	Dassie-rat (African rock rat)	Namibia, S Angola

Family Bathyergidae

African mole-rats; *c.* 11 species; Africa S of Sahara; desert, steppe, (forest); subterranean, feeding on roots, tubers, etc.

*Cape mole-rat
(Georychus capensis)*

Georychus

G. capensis	Cape mole-rat	S Africa; grassland, cultivation

Cryptomys; ref. 9.91.

C. damarensis	Damara mole-rat	Botswana, W Zambia etc.
C. hottentotus	Common mole-rat	Cape Prov. – S Zimbabwe; ? Zaire, Tanzania
C. mechowii		S Zaire, Zambia, etc.

C. natalensis	Natal mole-rat	Natal, Transvaal
C. ochraceocinereus	Ochre mole-rat	Ghana – N Uganda

Heliophobius

H. argenteocinereus	Silvery mole-rat	Zaire, Kenya – Zambezi
H. spalax	Thomas' silvery mole-rat	S Kenya

Bathyergus; dune mole-rats; S Africa; desert.

B. janetta	Namaqua dune mole-rat	Namaqualand, S Africa
B. suillus	Cape dune mole-rat	SW Cape Province, South Africa

Heterocephalus

H. glaber	Naked mole-rat	Ethiopia, Somalia, Kenya; steppe

Family Ctenodactylidae

Gundis; 5 species; Sahara, etc.; rocks in steppe and desert; herbivorous.

Ctenodactylus

C. gundi	Morocco – NW Libya
C. vali	S Morocco – Libya

*Gundi
(Ctenodactylus gundi)*

Pectinator

P. spekei	Somalia, Ethiopia; steppe

Massoutiera

M. mzabi	C Sahara

Felovia

F. vae	Senegal, Mauritania, Mali

ORDER LAGOMORPHA

Lagomorphs; *c.* 65 species; Eurasia, Africa, N, C, (S) America; desert – forest; terrestrial herbivores.

Family Ochotonidae

Pikas; *c.* 19 species; N, C Asia, W N America; grassland, rocks, mainly montane; herbivores.

*American pika
(Ochotona princeps)*

Ochotona

O. alpina		S Siberia, NE China, Hokkaido
O. collaris (*alpina*)	Collared pika	W Canada, Alaska
O. curzoniae	Black-lipped pika	Tibetan Plateau
O. daurica	Daurian pika	Altai – NE China
O. erythrotis	Chinese red pika	E Tibet, Qinghai, etc.
O. gaoligongensis		NW Yunnan, China; ref. 10.5

O. hyperborea (*alpina*)	Northern pika	NE Siberia
O. iliensis		Tien Shan, W China; ref. 10.6
O. kamensis	Kam pika	Kam, W Sichuan
O. koslowi	Kozlov's pika	N Tibet
O. ladacensis	Ladak pika	S Sinkiang, Kashmir; high montane
O. pallasii	Pallas's pika	Altai – Tien Shan
O. princeps (*alpina*)	American pika	Alaska – New Mexico
O. pusilla	Steppe pika	Volga – Irtysh; grassland
O. roylei (*nepalensis*) (*macrotis*)	Royle's pika (Large-eared pika)	N Burma – Himalayas – Tien Shan; high montane
O. rufescens	Afghan pika	Afghanistan, Iran, etc.
O. rutila	Turkestan red pika	Tien Shan – Pamirs; montane
O. thibetana	Moupin pika	W China, Tibet
O. thomasi	Thomas' pika	Qinghai, China

Family Leporidae

Rabbits, hares; *c.* 46 species; Eurasia, Africa, N, C, (S) America; (desert), steppe, savanna, (forest); terrestrial herbivores; ref. 10.1

Pentalagus

P. furnessi	Ryukyu rabbit	Ryukyu Is.; E

Brown hare
(*Lepus europaeus*)

Pronolagus; red rock rabbits, red hares; S, (E) Africa; savanna.

P. crassicaudatus	Greater red rock rabbit	E S Africa
P. randensis	Jameson's red rock rabbit	Zimbabwe – SW Africa
P. rupestris	Smith's red rock rabbit	S Africa – Kenya

Bunolagus; (*Lepus*).

B. monticularis	Riverine rabbit (Bushman hare)	Cape Province, S Africa; E

Romerolagus

R. diazi	Volcano rabbit (Zacatuche)	Mountains SE of Mexico City; E

Caprolagus

C. hispidus	Hispid hare	Nepal – Assam; E

Lepus; hares, jack rabbits; Eurasia, Africa, N America.

L. alleni	Antelope jack rabbit	NW Mexico, Arizona; steppe, desert
L. americanus	Snowshoe hare	Canada, N, W USA; open forest
L. arcticus (*timidus*)	American Arctic hare	Arctic America, Greenland

L. brachyurus	Japanese hare	Japan (except Hokkaido)
L. californicus (*insularis*)	Black-tailed jack rabbit	C, SW USA, N Mexico; grassland, steppe
L. callotis (*mexicanus*)	White-sided jack rabbit	S Mexico – S New Mexico
L. capensis (*tolai*) (? *granatensis*)	Cape hare	Africa, S Europe – C China
L. comus (*oiostolus*)	Yunnan hare	Yunnan, Guizhou, China; ref. 10.4
L. europaeus (*capensis*) (? *castroviejoi*)	Brown hare	W Europe – Baikal; [SC Canada, NC USA, Argentina, Chile]
L. fagani		Ethiopia, Sudan, ? Kenya; scrub
L. flavigularis	Tehuantepec jack rabbit	Oaxaca, S Mexico; E
L. habessinicus	Abyssinian hare	Ethiopia, N Somalia, Sudan
L. mandshuricus	Manchurian hare	Manchuria, etc.; forest
L. nigricollis	Indian hare	India, etc., Sri Lanka, [Java]
L. oiostolus	Woolly hare	Tibetan Plateau
L. peguensis (*siamensis*)	Burmese hare	Burma – Indochina, Hainan
L. saxatilis	Scrub hare	S Africa, Namibia; ref. 10.7
L. sinensis	Chinese hare	S China, Korea, Taiwan
L. starcki		Ethiopia; montane
L. timidus	Arctic hare (Mountain hare)	N Eurasia, arctic; tundra, open forest
L. townsendi	White-tailed jack rabbit	C, W USA; grassland, steppe
L. victoriae (*crawshayi*) (*whytei*)	Savanna hare	Transvaal – Somalia – Senegal; savanna; ref. 10.7
L. yarkandensis	Yarkand hare	SW Sinkiang; steppe

Poelagus

P. marjorita	Central African rabbit	Angola – Sudan; savanna

Sylvilagus; cottontails, American rabbits; N, C, (S) America.

S. aquaticus	Swamp rabbit	Texas – Georgia
S. audubonii	Desert cottontail	C, SW USA – C Mexico; desert, steppe
S. bachmani	Brush rabbit	W coast USA, Baja California; scrub
S. brasiliensis	Forest rabbit	NE Mexico – N Argentina; forest, scrub

S. cunicularius	Mexican cottontail	S Mexico
S. dicei		Costa Rica, W Panama
S. floridanus	Eastern cottontail	E, C USA – Costa Rica
S. graysoni	Tres Marias cottontail	Tres Marias Is, Mexico
S. insonus	Omilteme cottontail	Guerrero, Mexico; montane
S. nuttalli	Nuttall's cottontail	W USA; scrub, forest
S. palustris	Marsh rabbit	Florida – SE Virginia
S. transitionalis	New England cottontail	Appalachian Mts – S Maine

Oryctolagus

O. cuniculus	European rabbit	Iberia, NW Africa, [Europe, Australia, New Zealand, Chile]; grassland, open forest, cultivation; includes domestic rabbits

Sumatran rabbit
(Nesolagus netscheri)

Brachylagus; (*Sylvilagus*); ref. 10.2.

B. idahoensis	Pygmy rabbit	Montana, Oregon, Idaho, etc.; steppe, scrub

Nesolagus

N. netscheri	Sumatran rabbit	Sumatra; montane forest; I

ORDER MACROSCELIDEA

One family, sometimes included in order Insectivora.

Family Macroscelididae

Elephant-shrews; *c.* 15 species; Africa (except west); subdesert, steppe, savanna, forest; terrestrial; insectivores; ref. 10.3.

Rufous elephant-shrew
(Elephantulus rufescens)

Macroscelides

M. proboscideus	Short-eared elephant-shrew	SW, S Africa; steppe

Elephantulus; (*Nasilio*); small elephant-shrews; Africa; savanna – subdesert.

E. brachyrhynchus	Short-snouted elephant-shrew	E, S Africa
E. edwardii	Cape elephant-shrew	S Africa
E. fuscipes	Uganda elephant-shrew	Uganda, etc.
E. fuscus	Zambesi elephant-shrew (Peters' short-snouted elephant-shrew)	Lower Zambezi
E. intufi	Bushveld elephant-shrew	Namibia, etc.
E. myurus	Transvaal elephant-shrew	Zimbabwe – Cape Prov.
E. revoilii	Somalia elephant-shrew	Somalia
E. rozeti	North-African elephant-shrew	NW Africa

| *E. rufescens* | Rufous elephant-shrew | E Africa |
| *E. rupestris* | Rock elephant-shrew | SW, S Africa |

Petrodromus

| *P. tetradactylus* (*sultani*) (*tordayi*) | Four-toed elephant-shrew | E, SE, C Africa; savanna |

Rhynchocyon; (*Rhinonax*); forest elephant-shrews; E, C Africa; forest.

R. chrysopygus	Yellow-rumped elephant-shrew	Kenya coast
R. cirnei (*stuhlmanni*)	Chequered elephant-shrew	E, C Africa
R. petersi	Black and rufous elephant-shrew	SE Kenya, E Tanzania coast

Bibliographies

These bibliographies are arranged in three groups: general works on the diversity and classification of mammals; regional works, which are usually the best guides to identification; and works on particular groups of mammals. The first two are arranged in chronological order beginning with the most recent. The numbered taxonomic sources are arranged in a sequence corresponding to the main text and include all items referred to by number in the text. These include the original description of all species newly described in the last five years, along with the sources for classifications that do not follow those found in the general and regional works listed below.

General Works

IUCN 1988. *1988 IUCN red list of threatened animals*. Gland: IUCN, 154 pp.

Inskipp, T. & Bardzo, J. 1987. *World checklist of threatened mammals*. Peterborough: Nature Conservancy Council, 125 pp.

Anderson, S. & Jones, J.K. (Eds.) 1984. *Orders and families of recent mammals of the world*. New York etc.: Wiley, 686 pp.

Macdonald, D. (Ed.) 1984. *The encyclopaedia of mammals*, 2 vols. London: Allen & Unwin, 895 + xxxii pp.

Burton, J.A. 1984. Bibliography of Red Data Books. *Oryx*, **18**: 61–64.

Nowak, R.M. & Paradiso, J.L. 1983. *Walker's mammals of the world*, 4th ed. Baltimore & London: Johns Hopkins University Press, 1362 pp. (2 vols.)

Honacki, J.H., Kinman, K.E. & Koeppl, J.W. (Eds.) 1982. *Mammal species of the world*. Lawrence, Kansas: Allen Press, ix + 694 pp.

Thornback, J. & Jenkins, M. 1982. *The IUCN mammal red data book. Part 1. Threatened mammalian taxa of the Americas and the Australian zoogeographic region (excluding Cetacea)*. Gland, Switzerland: IUCN, 516 pp.

Hickman, G.C. 1981. National mammal guides: a review of references to Recent faunas. *Mammal Review*, **11**: 53–85.

McKenna, M.C. 1975. Towards a phylogenetic classification of the Mammalia. *In* Luckett, W.P. & Szaley, F.S. (Eds.) *Phylogeny of the Primates*. New York: Plenum Publishing Co.: 21–46. (A provisional classification of fossil and recent mammals to ordinal level.)

Anderson, S. (Ed.) 1969–. *Mammalian species*. American Society of Mammalogists. (A loose-leaf part-work, each part dealing with a single species in 2–8 pages.)

Simpson, G.G. 1945. The principles of classification and a classification of the mammals. *Bulletin of the American Museum of Natural History*, **85**: i–xvi, 1–350. (Lists all recent and fossil genera.)

Geographical sources

In the following lists priority has been given to selecting reference works that are recent, authoritative with respect to taxonomy, comprehensive in coverage of species, useful for identification and that cover a wide area and include distribution maps, although few meet all of these criteria.

Palaearctic Region (Europe, N. Asia, N. Africa)

Aulagnier, S. & Thevenot, M. 1986. *Catalogue des mammifères sauvages du Maroc*. Rabat: Institut Scientifique Charia Ibn Batouta BP, 163 pp.

Corbet, G.B. 1984. *The mammals of the Palaearctic Region: a taxonomic review: supplement*. London: British Museum (Natural History), 45 pp.

Niethammer, J. & Krapp, F. 1978, 1982. *Handbuch der Säugetiere Europas*. Vols. 1 and 2 (1): Rodentia. Wiesbaden: Akademische Verlagsgesellschaft, 476 + 649 pp.

Pucek, Z. (Ed.) 1981. *Keys to the vertebrates of Poland: mammals*. Warsaw: translated and published for Smithsonian Institution and National Science Foundation, Washington DC by Polish Scientific Publishers, 367 pp.

Gromov, I.M. & Baranova, G. 1981. [*Catalogue of mammals of the USSR: Pliocene to the present day*]. Leningrad: Akademia Nauk SSSR. 455 pp.

Corbet, G.B. & Ovenden, D. 1980. *The mammals of Britain and Europe.* London: Collins, 253 pp.

Sokolov, V.E. & Orlov, V.N. 1980. [*Key to the mammals of the Mongolian People's Republic.*] Moscow, 352 pp. (In Russian).

Corbet, G.B. 1978. *The mammals of the Palaearctic Region: a taxonomic review.* London: British Museum (Natural History), 314 pp. (Includes keys and distribution maps for all species.)

Corbet, G.B. & Southern, H.N. (Eds.) 1977. *The handbook of British mammals,* 2nd ed. Oxford etc: Blackwell, 520 pp.

Saint Girons, M.-C. 1973. *Les mammifères de France et du Benelux.* Paris: Doin, 481 pp.

Bobrinski, N.A., Kuznetsov, B.A. & Kuzyakin, A.P. 1965 *Opredelitel' mlekopitayushchikh SSSR.* [Key to mammals of the USSR]. Moscow. (Coloured illustrations, text in Russian.)

Harrison, D.L. 1964–72. *The mammals of Arabia.* London: Ernest Benn, 3 vols.

Imaizumi, Y. 1960. *Coloured illustrations of the mammals of Japan.* Osaka: Hoikusha Publishing Co., 196 pp. (Text in Japanese.)

Shou, Z.-H. 1964. [*Handbook of economic animals of China. Mammals*]. Peking, 554 pp, 72 pls.

Australasian Region

Bannister, J.L. *et al.* 1988. *Zoological catalogue of Australia. Vol. 5. Mammalia.* Canberra: Australian Government Publishing Service.

Strahan, R. (Ed.) 1983. *Complete book of Australian mammals.* London etc: Angus & Robertson, 530 pp.

Ziegler, A.C. 1982. An ecological check-list of New Guinea recent mammals, pp. 863–894. *In* Gressitt, J.L. (Ed.) *Biogeography and ecology of New Guinea,* vol. 2. Monographiae Biologicae, vol. 42. The Hague: W. Junk, 983 pp.

Tyler, M.J. 1979. *The status of endangered Australian wildlife.* Adelaide: Royal Zoological Society of South Australia, 210 pp.

Oriental Region (SE Asia)

Musser, G.G. 1987. The mammals of Sulawesi. *In* Whitmore, T.C. (ed.) *Biogeographical evolution of the Malay archipelago.* Oxford: Clarendon Press, 73–93.

Heaney, R.L. *et al.* 1987. An annotated checklist of the taxonomic and conservation status of land mammals in the Philippines. *Silliman Journal,* 34: 32–66.

Medway, Lord. 1983. *The wild mammals of Malaya* . . . . 2nd ed. (revised). Kuala Lumpur etc: Oxford University Press.

Phillips, W.W.A. 1980–84. *Manual of the mammals of Sri Lanka,* 2nd ed. In 3 parts. Sri Lanka: Wildlife and Nature Protection Society, 390 pp.

Medway, Lord 1977. *Mammals of Borneo: field keys and an annotated checklist.* Kuala Lumpur: Malaysian Branch Royal Asiatic Society (monograph no.7) xii + 172 pp.

Lekagul, B. & McNeely, J.A. 1977. *Mammals of Thailand.* Bangkok. (All species, with illustrations, keys and maps.)

Roberts, T.J. 1977. *The mammals of Pakistan.* London: Ernest Benn, 361 pp. (Includes keys, illustrations and maps.)

Eisenberg, J.F. & McKay, G.M. 1970. An annotated checklist of the mammals of Ceylon with keys to the species. *Ceylon Journal of Science (Biol.),* **8**: 69–99.

Van Peenen, P.F.D., Ryan, P.F. & Light, R.H. 1969. *Preliminary identification manual for mammals of South Vietnam.* Washington: Smithsonian Institution, 310 pp.

Ellerman, J.R. & Morrison-Scott, T.C.S. 1951. *Checklist of Palaearctic and Indian mammals 1758–1946.* London: British Museum (Natural History), 810 pp. (Covers Oriental Region north of 10°N in Malaya.)

Africa

Ansell, W.F.H. & Dowsett, R.J. 1988. *Mammals of Malawi/ an annotated checklist.* Zennor: Trendrine Press, 170 pp.

Meester, J.A.J. *et al.* 1986. *Classification of southern African mammals.* Transvaal Museum Monograph, no. 5: 359 pp.

Smithers, R.H.N. 1983. *The mammals of the southern African subregion.* Pretoria: University of Pretoria, xxii + 736 pp.

Kingdon, J. 1971–82. *East African mammals: an atlas of evolution in Africa*, 7 vols. London etc: Academic Press.

Haltenorth, T. & Diller, H. 1980. *A field guide to the mammals of Africa including Madagascar*. London: Collins, 400 pp.

Osborn, D.J. & Helmy, I. The contemporary land mammals of Egypt (including Sinai). *Fieldiana: Zoology*, N.S. **5**: xix + 579 pp.

Swanepoel, P., Smithers, R.H.N. & Rautenbach, I.L. 1980. A checklist and numbering system of the extant mammals of the southern African subregion. *Annals of the Transvaal Museum*, **32**: 155–196.

Ansell, W.F.H. 1978. *The mammals of Zambia*. Chilanga: Department of National Parks & Wildlife Service, 126 pp. + 204 maps.

Largen, M.J., Kock, D. & Yalden, D.W. 1974. Catalogue of the mammals of Ethiopia. 1. Chiroptera. *Monitore Zoologico Italiano. (N.S.)*, Supplement, **5**: 221–298.

Yalden, D.W., Largen, M.J. & Kock, D. 1976. Catalogue of the mammals of Ethiopia. 2. Insectivora and Rodentia. *Ibid*. Supplement, **8**: 1–118; 1977. 3. Primates. *Ibid*. Supplement, **9**: 1–52; 1980. 4. Carnivora. *Ibid*. Supplement, **13**: 169–272. 1984. 5. Artiodactyla. *Ibid*. Supplement, **19**: 67–221. 1986. 6. Perissodactyla-Cetacea. *Ibid*. Supplement, **21**: 31–103.

Meester, J. & Setzer, H.W. (Eds.) 1971. *The mammals of Africa: an identification manual*. Washingtion: Smithsonian Institution. (Includes keys to all species.)

Dorst, J. & Dandelot, P. 1970. *A field guide to the larger mammals of Africa*. London: Collins, 287 pp.

North and Central America, West Indies

Jones, J.K. *et al.* 1986. Revised checklist of North American mammals north of Mexico. *Occasional Papers, Museum Texas Tech University*, **107**: 22 pp.

Zyll de Jong, C.G. van 1983. *Handbook of Canadian mammals. I. Marsupials and insectivores*. Ottawa: National Museum of Natural Sciences, 210 pp.

Wilson, D.E. 1983. Checklist of mammals. Pp. 443–447 in Janzen, D.H. (Ed.) *Costa Rica natural history*. Chicago: University of Chicago Press, 816 pp.

Chapman, J.A. & Feldhammer, G.A. 1982. *Wild mammals of North America: biology, management and economics*. Baltimore etc: Johns Hopkins Press, 1184 pp.

Pulido, J.R. *et al.* 1982. *Catalogo de los mamiferos terrestres nativos de Mexico*. Mexico: Ed. Trillas, 126 pp.

Hall, E.R. 1981. *The mammals of North America*, 2nd edn. New York: Wiley, xv + 1181 + 90 pp, 2 vols. (Includes Central America south to Panama and the West Indies.)

Banfield, A.W.F. 1974. *The mammals of Canada*. Toronto.

South America

Eisenberg, J. 1989. *Mammals of the Neotropics. Vol. 1: the northern neotropics . . .* Chicago: Univ. Chicago Press.

Husson, A.M. 1978. *The mammals of Surinam*. Leiden: Brill, xxxiv + 569 pp, 161 pls.

Peterson, N.E. & Pine, R.H. 1982. [Key for identification of the mammals of the Brazilian Amazon region with the exception of bats and primates]. *Acta Amazonica*, **12**: 465–482. (Portuguese, English summary.)

Orlog, C.C. & Lucero, M.M. 1981. *Guia de los mammiferos Argentinos*. Tucuman, 151 pp.

Mann Fischer, G. 1978. Los pequinos mamiferos de Chile (marsupiales, quiropteros, edentados y roedores). *Guyana* (Zool.), **40**: 342 pp.

Handley, C.O. 1976. Mammals of the Smithsonian Venezuelan Project. *Brigham Young University Science Bulletin*, Biol. Ser. **20** (5): 1–89. (All species recorded from Venezuela are mentioned.)

Ximenez, A., Langguth, A. & Praderi, R. 1972. Lista sistematica de los mamiferos del Uruguay. *Anales del Museo de Historia Natural de Montevideo*, 7 (5): 1–49.

Cabrera, A. 1957, 1961. Catalogo de los mamiferos de America del Sur. *Revista del Museo Argentino de Ciencias Naturales 'Bernardino Rivadavia'*, Ser. Cienc. zool. **4**, 732 pp. (A detailed nomenclatorial checklist.)

Taxonomic sources

The following sources have been used to update the basic geographical sources in compiling the list. They are quoted in the text of the list and are arranged here approximately in the sequence in which they appear in the list.

1. Monotremata, Marsupialia

1.1 Kirsch, J.A.W. & Calaby, J.H. 1977. The species of living marsupials: an annotated list. *In* Stonehouse, B. & Gilmore, D. (eds.) *The biology of marsupials.* London etc.: Macmillan: 9–26.

1.2 Archer, M. 1982. *Carnivorous marsupials.* 2 vols. Mosman, NSW, Australia: Royal Zoological Society of NSW, 804 pp.

1.3 Pine, R.H. & Handley, C.O. 1984. A review of the Amazonian short-tailed opossum *Monodelphis emiliae* (Thomas). *Mammalia,* **48**: 239–245.

1.4 Massoia, E. 1980. Un marsupial nuevo para la Argentina: *Monodelphis scalops* ... *Physis,* **39,** C: 61–66.

1.5 Kitchener, D.J., Stoddart, J. & Henry, J. 1983. A taxonomic appraisal of the genus *Ningaui* Archer ... including description of a new species. *Australian Journal of Zoology,* **31**: 361–379.

1.6 Kitchener, D.J., Stoddart, J. & Henry, J. 1984. A taxonomic revision of the *Sminthopsis murina* complex ... in Australia ... *Record of the Western Australian Museum,* **11**: 201–248.

1.7 McKenzie, N.L. & Archer, M. 1982. *Sminthopsis youngsoni* (Marsupialia: Dasyuridae) the lesser hairy-footed dunnart, a new species from arid Australia. *Australian Mammalogy,* **5** (4): 267–279.

1.8 Lyne, A.G. & Mort, P.A. 1981. A comparison of skull morphology in the marsupial bandicoot genus *Isoodon* ... *Australian Mammalogy,* **4**: 107–133.

1.9 McKay, G.M. 1982. Nomenclature of the gliding possum genera *Petaurus* and *Petauroides* ... *Australian Mammalogy,* **5**: 37–39.

1.10 Maynes, G.M. 1982. A new species of rock wallaby, *Petrogale persephone* ... from Prosperine, central Queensland. *Australian Mammalogy,* **5**: 47–58.

1.11 Groves, C.P. 1982. The systematics of tree kangaroos (*Dendrolagus* ...). *Australian Mammalogy,* **5**: 157–186.

1.12 Pine, R.H. 1981. Review of the mouse opossums *Marmosa parvidens* Tate and *Marmosa invicta* Goldman ... *Mammalia,* **45**: 55–70.

1.13 Van Dyck, 1980. The cinnamon antechinus, *Antechinus leo* ... a new species from the vine-forests of Cape York Peninsula. *Australian Mammalogy,* **3**: 5–17.

1.14 Ziegler, A.C. 1981. *Petaurus abidi,* new species of glider ... from Papua New Guinea. *Australian Mammalogy,* **4**: 81–88.

1.15 Seebeck, J.H. & Johnston, P.G. 1980. *Potorous longipes,* new species ... from eastern Victoria, Australia. *Australian Journal of Zoology,* **28**: 119–134.

1.16 Archer, M.A. 1982. Review of the dasyurid (Marsupialia) fossil record, ... Pp. 397–443 *in* Archer, M. (Ed.) *Carnivorous marsupials.* Mosman, New South Wales: Royal Zoological Society of New South Wales.

1.17 Creighton, G.K. 1985. Phylogenetic inference ... in *Marmosa. Acta Zoologica Fennica,* **170**: 121–124.

1.18 Flannery, T. 1987. A new species of *Phalanger* ... from western Papua New Guinea. *Records of the Western Australian Museum,* **39**: 183–194.

1.19 Menzies, J.I. & Pernetta, J.C. 1986. A taxonomic revision of cuscuses allied to *Phalanger orientalis* ... *Journal of Zoology, London,* **(B)**: 551–618.

1.20 Kitchener, D.J. 1988. A new species of false antechinus ... from Kimberley, Western Australia. *Records of the Western Australian Museum,* **14**: 61–71.

1.21 Kitchener, D.J. & Caputi, N. 1988. A new species of false antechinus ... from Western Australia. *Records of the Western Australian Museum* **14**: 35–59.

1.22 Van Dyck, S. 1988. The bronze quoll, *Dasyurus sparticus,* new species ... from the savannas of Papua New Guinea. *Australian Mammalogy* **11**: 145–156.

1.23 Van Dyck, S. 1985. *Sminthopsis leucopus* ... in North Queensland. *Australian Mammalogy* **8**: 53–60.

1.24 Van Dyck, S. 1986. The chestnut dunnart, *Sminthopsis archeri*, new species . . . *Australian Mammology* **9**: 111–124.

1.25 Bublitz, J. 1987. Untersuchungen zur Systematik der rezenten Caenolestidae . . . *Bonner Zoologische Monographien* no. 23: 96 pp.

1.26 Flannery, T. *et al.* 1987. The phylogenetic relationships of living phalangerids . . . Pp. 477–506 in Archer, M. (ed.) *Possums and opossums.* Chipping Norton, New South Wales: Beatty, 2 vols, 788 pp.

1.27 Reig, O.A. *et al.* 1985. New conclusions on the relationships of the opossum-like marsupials . . . *Ameghiniana* **21**: 335–343.

1.28 Gardner, A.L. & Creighton, G.K. 1989. A new generic name for Tate's (1933) *microtarsus* group of South America mouse opossums . . . *Proceeding of the Biological Society of Washington* **102**: 3–7.

1.29 Vivo, M. de & Gomes, N.F. 1989. First record of *Caluromysiops irrupta* . . . from Brasil. *Mammalia,* **62**: 310–311.

2. Edentata

2.1 Montgomery, G.G. (ed.) 1985. *The evolution and ecology of armadillos, sloths, and vermilinguas.* Washington: Smithsonian Institution Press, 451 pp.

2.2 Wetzel, R.M. 1985. The identification and distribution of Recent Xenarthra (=Edentata). Pp. 5–21 in Montgomery, G.G. (above).

2.3 Wetzel, R.M. 1985. Taxonomy and distribution of armadillos, Dasypodidae. Pp. 23–46 in Montgomery, G.G. (above).

3. Insectivora

3.1 Poduschka, W. & Poduschka, C. 1982. Die taxonomische Zugehorigkeit von *Dasogale fontoynonti* G. Grandidier, 1928. *Sitzungsberichte der Osterreichischen Akademie der Wissenschaften. Mathematisch-Naturwissenschaftliche Klasse,* **191**: 253–264.

3.2 MacPhee, R.D.E. 1987. The shrew tenrecs of Madagascar . . . *American Museum Novitates,* **2889**: 45 pp.

3.3 Heaney, C.R. & Morgan, G.S. 1982. A new species of gymnure, *Podogymnura,* (Mammalia, Erinaceidae) from Dinagat Island, Philippines. *Proceedings of the Biological Society of Washington,* **95**: 13–26.

3.4 Junge, J.A. & Hoffmann, R.S. 1981. An annotated key to the long-tailed shrews (genus *Sorex*) of the United States and Canada, with notes on Middle American *Sorex. Occasional Papers of the Museum of Natural History, University of Kansas, Lawrence,* **94**: 48 pp.

3.5 Hoffmann, R.S. 1986. A review of the genus *Soriculus* . . . . *Journal Bombay Natural History Society,* **82**, 459–481.

3.6 Jenkins, P.D. 1982. A discussion of Malayan and Indonesian shrews of the genus *Crocidura* . . . *Zoologische Mededeelingen,* **56**: 267–279.

3.7 Hutterer, R. 1981. *Crocidura manengubae* n. sp. . . . . eine neue Spitzmaus aus Kamerun. *Bonner Zoologische Beitrage,* **32**: 241–248.

3.8 Hutterer, R. 1983. Status of some African *Crocidura* described by Isidore Geoffroy Saint-Hilaire, Carl J. Sundevall and Theodor von Heuglin. *Annales, Musée Royal de l'Afrique Centrale, Sciences Zoologiques,* **237**: 207–217.

3.9 Hutterer, R. 1983. Taxonomy and distribution of *Crocidura fuscomurina* (Heuglin, 1863). *Mammalia,* **47**: 221–227.

3.10 Hutterer, R. & Kock, D. 1983. Spitzmäuse aus den Nuba-Bergen Kordofans, Sudan. *Senckenbergiana Biologica,* **63**: 17–26.

3.11 Hutterer, R. 1983. *Crocidura grandiceps,* eine neue Spitzmaus aus Westafrika. *Revue Suisse de Zoologie,* **90**: 699–707.

3.12 Hutterer, R. & Happold, D.C.D. 1983. The shrews of Nigeria . . . *Bonner Zoologische Monographien,* **18**: 79 pp.

3.13 Jenkins, P.D. 1984. Description of a new species of *Sylvisorex* . . . from Tanzania. *Bulletin of the British Museum* (*Natural History*) (*Zoology*), **47**: 65–76.

3.14 Hoffmann, R.S. 1984. A review of the shrew-moles (genus *Uropsilus*) of China and Burma. *Journal of the Mammalogical Society of Japan,* **10**: 69–80.

3.15 Stogov, I.I. 1985. On two little studied species of white-toothed shrews . . . from the mountain regions in the south of the USSR. *Zoologicheskii Zhurnal,* **64**: 264–268.

3.16 Dippenaar, N.J. 1980. New species of *Crocidura* from Ethiopia and Northern Tanzania . . . *Annals of the Transvaal Museum,* **32**: 125–154.

3.17 Jenkins, P.D. 1988. A new species of *Microgale* . . . *American Museum Novitates,* **2910**; 7 pp.

3.18 Corbet, G.B. 1988. The family Erinaceidae . . . *Mammal Review,* **18**: 117–172.

3.19 Hoffmann, R.S. 1987. Review of . . . Chinese red-toothed shrews . . . *Acta Theriologica Sinica,* **7**: 100–139.

3.20 Zaitsev, M.V. 1988. On the nomenclature of the red-toothed shrews of the genus *Sorex* in the fauna of the USSR. *Zoologicheskii Zhurnal,* **67**: 1878–1888.

3.21 Yoshiyuki, M. & Imaizumi, Y. 1986. A new species of *Sorex* from Sado Island, Japan. *Bulletin of the National Museum Tokyo,* ser. A, **12**: 185–193.

3.22 Hutterer, R. & Dippenaar, N.J. 1987. A new species of *Crocidura* from Zambia. *Bonner Zoologische Beiträge,* **38**: 1–7.

3.23 Hutterer, R. & Dippenaar, N.J. 1987. *Crocidura ansellorum,* emended species name . . . *Bonner Zoologische Beiträge,* **38**: 269.

3.24 Hutterer, R. & Harrison, D.L. 1988. A new look at the shrews . . . of Arabia. *Bonner Zoologische Beiträge,* **39**: 59–72.

3.25 Hutterer, R. *et al.* 1987. The shrews of the eastern Canary Islands . . . *Journal of Natural History,* **21**: 1347–1357.

3.26 Vogel, P. *et al.* 1989. The correct name, species diagnosis, and distribution of the Sicilian shrew. *Bonner Zoologische Beiträge* **40**: 243–248.

3.27 Dippenaar, N.J. & Meester, J.A.J. 1989. Revision of the *luna-fumosa* complex of Afrotropical *Crocidura. Annals of the Transvaal Museum,* **35**: 1–47.

3.28 Hutterer, R. 1986. The species of *Crocidura* . . . in Morocco. *Mammalia,* **50**: 521–534.

3.29 Hutterer, R. 1986. Eine neue Soricidengattung aus Zentralafrika . . . *Zeitschrift für Säugetierkunde,* **51**: 257–266.

3.30 Hutterer, R. 1986. Synopsis der Gattung *Paracrocidura* . . . *Bonner Zoologische Beiträge,* **37**: 73–90.

3.31 Hutterer, R. 1986. Diagnosen neuer Spitzmäuse aus Tanzania . . . *Bonner Zoologische Beiträge,* **37**: 23–33.

3.32 Feiler, A. 1988. Die Säugetiere der Inseln im Golf von Guinea . . . *Zoologische Abhandlungen staatliches Museum für Tierkunde Dresden,* **44**: 83–88.

3.33 Vogel, P. *et al.* 1986. A contribution to the taxonomy and ecology of shrews . . . from Crete and Turkey. *Acta Theriologica,* **31**: 537–545.

3.34 Hutterer, R. & Verheyen, W. 1985. A new species of shrew, genus *Sylvisorex,* from Ruanda and Zaire . . . *Zeitschrift für Säugetierkunde,* **50**: 266–271.

3.35 Molina, O.M. & Hutterer, R. 1989. A cryptic new species of *Crocidura* from Gran Canaria and Tenerife, Canary Islands. *Bonner Zoologische Beiträge,* **40**: 85–97.

4. Chiroptera

4.1 Miller, G.S. 1907. The families and genera of bats. *Bulletin of the United States National Museum,* **57**: i–xvii, 1–282, 49 pls.

4.2 Yoshiyuki, M. 1989. *A systematic study of the Japanese Chiroptera.* National Science Museum, Tokyo, i–vi, 1–242.

4.3 Barbour, R.W. & Davis, W.H. 1969. *Bats of America.* Lexington: University Press of Kentucky.

4.4 Villa-R, B. 1966. *Los murcielagos de Mexico.* Mexico: Instituto de Biologia, Universidad Nacional Autónoma de Mexico.

4.5 Husson, A.M. 1962. The bats of Suriname. *Zoologische Verhandelingen, Leiden,* no. 58, 282 pp., 30 pls.

4.6 Goodwin, G.G. & Greenhall, A.M. 1961. A review of the bats of Trinidad and Tobago. *Bulletin of the American Museum of Natural History,* **122**: 187–302, 40 pls.

4.7 Baker, R.J. & Genoways, H.H. 1978. Zoogeography of Antillean bats. *In* Zoogeography in the Caribbean. *Special Publications. Academy of Natural Sciences of Philadelphia,* **13**: 53–97.

4.8 Brosset, A. 1959. The bats of Central and Western India. *Journal of the Bombay Natural History Society,* **59**: 1–57, 583–624, 707–746, 8 pls.

4.9   Andersen, K. 1912. *Catalogue of the Chiroptera in the collection of the British Museum. 1 Megachiroptera.* 2nd ed. London: British Museum (Natural History).

4.10   Koopman, K.F. 1980. Zoogeography of mammals from islands off the northern coast of New Guinea. *American Museum Novitates,* no. 2690: 17 pp.

4.11   Rookmaaker, L.C. & Bergmans, W. 1981. Taxonomy and geography of *Rousettus amplexicaudatus* (Geoffroy, 1810) . . . *Beaufortia,* **31**: 1–29.

4.12   Hill, J.E. 1983. Bats . . . from Indo-Australia. *Bulletin of the British Museum (Natural History),* (Zool.), **45**: 103–208.

4.13   Bergmans, W & Hill, J.E. 1980. On a new species of *Rousettus* Gray, 1821, from Sumatra and Borneo . . . *Bulletin of the British Museum (Natural History),* (Zool.), **38**: 95–104.

4.14   Hill, J.E. & Francis, C.M. 1984. New bats . . . and new records of bats from Borneo and Malaya. *Bulletin of the British Museum (Natural History),* (Zool.), **47**: 305–329.

4.15   Bergmans, W. 1980. A new fruit bat of the genus *Myonycteris* Matschie, 1899, from eastern Kenya and Tanzania . . . *Zoologische Mededeelingen, Leiden,* **55**: 171–181.

4.16   Musser, G.G., Koopman, K.F. & Cafifia, D. 1982. The Sulawesian *Pteropus arquatus* and *P. argentatus* are *Acerodon celebensis*; the Philippine *P. leucotis* is an *Acerodon. Journal of Mammalogy,* **63**: 319–328.

4.17   Cheke, A.S. & Dahl, A.F. 1981. The status of bats on western Indian Ocean Islands, with special reference to *Pteropus. Mammalia,* **45**: 205–238.

4.18   Klingener, D. & Creighton, G.K. 1984. On small bats of the genus *Pteropus* from the Philippines. *Proceedings of the Biological Society of Washington,* **97**: 395–403.

4.19   Bergmans, W. & Sarbini, S. 1985. Fruit bats of the genus *Dobsonia* Palmer, 1898 from the islands of Biak, Owii, Numfoor and Yapen, Irian Jaya . . . *Beaufortia,* **34**: 181–189.

4.20   Yenbutra, S. & Felten, H. 1983. A new species of the fruit bat genus *Megaerops* from SE-Asia. *Senckenbergiana Biologica,* **64**: 1–11.

4.21   Smith, J.D. & Hood, C.S. 1983. A new species of tube-nosed fruit bat (*Nyctimene*) from the Bismarck Archipelago, Papua New Guinea. *Occasional Papers. The Museum, Texas Tech University,* no. **81**: 1–14.

4.22   Heaney, L.R. & Peterson, R.L. 1984. A new species of tube-nosed fruit bat (*Nyctimene*) from Negros island, Philippines . . . *Occasional Papers of the Museum of Zoology, University of Michigan,* no. **708**: 1–16.

4.23   Rozendaal, F.G. 1984. Notes on macroglossine bats from Sulawesi and the Moluccas, Indonesia . . . *Zoologische Mededeelingen, Leiden,* **58**: 187–212.

4.24   Goodwin, R.E. 1978. The bats of Timor: systematics and ecology. *Bulletin of the American Museum of Natural History,* **163**: 73–122.

4.25   Ziegler, A.C. 1982. The Australo-Papuan genus *Syconycteris* . . . *Occasional Papers of the Bernice Pauahi Bishop Museum,* **25** (5): 1–22.

4.26   Smith, J.D. & Hood, C.S. 1981. Preliminary notes on bats from the Bismarck Archipelago . . . *Science in New Guinea,* **8**: 81–121.

4.27   Kitchener, D.J. 1980. *Taphozous hilli* sp. nov. . . . a new sheath-tailed bat from Western Australia and Northern Territory. *Records of the Western Australian Museum,* **8**: 161–169.

4.28   McKean, J.L. & Friend, G.R. 1979. *Taphozous kapalgensis,* a new species of sheath-tailed bat from the Northern Territory, Australia. *Victoria Naturalist,* **96**: 239–241.

4.29   Kock, D. 1981. Zwei Fledermäuse neu für Kenya . . . *Senckenbergiana Biologica,* (1980), **61**: 321–327.

4.30   Bohme, W. & Hutterer, R. 1978. Kommentierte Liste einer Säugetier-Aufsammlung aus dem Senegal. *Bonner Zoologische Beitrage,* **29**: 303–322.

4.31   Hill, J.E. & Yoshiyuki, M. 1980. A new species of *Rhinolophus* . . . from Iriomote Island, Ryukyu Islands . . . *Bulletin of the National Science Museum, Tokyo,* Ser. A. (Zool.), **6**: 179–189.

4.32   Smith, J.D. & Hill, J.E. 1981. A new species and subspecies of bat of the *Hipposideros bicolor* – group from Papua New Guinea . . . *Contributions in Science,* **331**: 1–19.

4.33   Jenkins, P.D. & Hill, J.E. 1981. The status of *Hipposideros galeritus* Cantor, 1846 and *Hipposideros cervinus* (Gould, 1854) . . . *Bulletin of the British Museum (Natural History),* (Zool.), **41**: 279–294.

4.34   Khajuria, H. 1970. A new leaf-nosed bat from Central India. *Mammalia,* **34**: 622–627.

4.35 Khajuria, H. 1982. External genitalia and bacula of some central Indian Microchiroptera. *Säugetierkundliche Mitteilungen*, **30**: 287–295.

4.36 Hill, J.E. & Yenbutra, S. A new species of the *Hipposideros bicolor* group . . . from Thailand. *Bulletin of the British Museum (Natural History)*, (Zool.), **47**: 77–82.

4.37 Brosset, A. 1984. Chiroptères d'altitude du Mont Nimba (Guinée). Description d'une espèce nouvelle, *Hipposideros lamottei*. *Mammalia*, **48**: 544–555.

4.38 Hill, J.E. 1982. A review of the leaf-nosed bats *Rhinonycteris, Cloeotis* and *Triaenops*. *Bonner Zoologische Beitrage*, **33**: 165–186.

4.39 Silva Taboada, G. 1976. Historia y actualizatión taxónomica de algunes especies Antillanas de murciélagos de los géneros *Pteronotus, Brachyphylla, Lasiurus*, y *Antrozous* . . . *Poeyana*, **153**: 1–24.

4.40 Hall, E.R. 1981. *The mammals of North America*. New York, etc: Wiley.

4.41 Davis, W.B. 1976. Notes on the bats *Saccopteryx canescens* Thomas and *Micronycteris hirsuta* (Peters). *Journal of Mammalogy*, **57**: 604–607.

4.42 Swanepoel, P. & Genoways, H.H. 1979. Morphometrics. *In* Baker, R.J., Jones, J.K. & Carter, D.C. (eds.) Biology of bats of the New World family Phyllostomatidae. Part III. *Special Publications. The Museum, Texas Tech University*, **16**: 13–106.

4.43 Williams, S.L. & Genoways, H.H. 1980. Results of the Alcoa Foundation – Suriname Expeditions. II. Additional records of bats . . . from Suriname. *Annals of the Carnegie Museum*, **49**: 213–236.

4.44 Ochoa, J. & Ibanez, C. 1984. Nuevo murcielago del genero *Lonchorhina* . . . *Memorias de la Sociedad de Ciencias Naturales 'La Salle'* (1982), **42** (118): 145–159.

4.45 Gardner, A.L. 1976. The distributional status of some Peruvian mammals. *Occasional Papers of the Museum of Zoology, Louisiana State University*, **48**: 1–18.

4.46 Genoways, H.H. & Williams, S.L. 1980. Results of the Alcoa Foundation – Suriname Expeditions. I. A new species of bat of the genus *Tonatia*. *Annals of the Carnegie Museum*, **49**: 203–211.

4.47 Webster, W.D. & Jones, J.K. 1980. Taxonomic and nomenclatorial notes on bats of the genus *Glossophaga* in North America, with description of a new species. *Occasional Papers. The Museum, Texas Tech University*, **71**: 1–12.

4.48 Taddei, V.A., Vizotto, L.D. & Sazima, I. 1982. Una nova espéce de *Lonchophylla* do Brasil e clave para identifição das espécies do gênero . . . *Ciencia et Cultura*, **35**: 625–629.

4.49 Hill, J.E. 1980. A note on *Lonchophylla* . . . from Ecuador and Peru, with the description of a new species. *Bulletin of the British Museum (Natural History)*, (Zool.), **38**: 233–236.

4.50 Handley, C. 1984. New species of mammals from northern South America: a long-tongued bat, genus *Anoura* Gray. *Proceedings of the Biological Society of Washington*, **97**: 513–521.

4.51 Davis, W.B. 1980. New *Sturnira* .... from Central and South America with key to currently recognised species. *Occasional Papers. The Museum, Texas Tech University*, no. 70: 1–5.

4.52 Anderson, S., Koopman, K.F. & Creighton, G.K. 1982. Bats of Bolivia: an annotated checklist. *American Museum Novitates*, no. 2750: 1–24.

4.53 Handley, C.D. 1976. Mammals of the Smithsonian Venezuelan Project. *Brigham Young University Science Bulletin*, Biol. Ser. 20, (5): 1–89, map.

4.54 Greenbaum, I.F., Baker, R.J. & Wilson, D.E. 1975. Evolutionary implications of the karyotypes of the stenodermine genera *Ardops, Ariteus, Phyllops*, and *Ectophylla*. *Bulletin of the Southern California Academy of Sciences*, **74**: 156–159.

4.55 Jones, J.K. & Carter, D.C. 1979. Systematic and distributional notes. *In* Baker, R.J., Jones, J.K. & Carter, D.C. (Eds.) Biology of bats of the New World family Phyllostomatidae. Part III. *Special Publications. The Museum, Texas Tech University*, no. 16: 7–11.

4.56 Koopman, K.F. 1978. Zoogeography of Peruvian bats with special emphasis on the role of the Andes. *American Museum Novitates*, no. 2651: 1–33.

4.57 Davis, W.B. 1984. Review of the large fruit-eating bats of the *Artibeus 'lituratus'* complex . . . in Middle America. *Occasional Papers. The Museum, Texas Tech University*, no. 93: 1–16.

4.58 Swanepoel, P. & Genoways, H.H. 1978. Revision of the Antillean bats of the genus

Brachyphylla . . . *Bulletin of Carnegie Museum of Natural History,* no. 12, 53 pp.

4.59 Koopman, K.F. 1982. In Honacki, J.H., Kinman, K.E. & Koeppl, J.W. *Mammal species of the world. A taxonomic and geographical reference.* Lawrence, Kansas: Allen Press Inc./Association of Systematics Collections.

4.60 Ruschi, A. 1951. Morcegos do Estado do Espirito Santo. Familia Vespertilionidae, . . . *Boletim do Museu de Biologia 'Prof. Mello-Leitao'.* (Zool.), no. 4: 1–11.

4.61 Pine, R.H. & Ruschi, A. 1976. Concerning certain bats described and recorded from Espirito Santo, Brazil. *Anales del Instituto de Biologia, Universidad de Mexico,* **47,** Ser. Zool. (2): 183–196.

4.62 Horácek, I. & Hanák, V. 1984. Comments on the systematics and phylogeny of *Myotis nattereri* (Kuhl, 1818). *Myotis,* **21–22:** 20–29.

4.63 Bogan, M.A. 1978. A new species of *Myotis* from the Islas Tres Marias, Nayarit, Mexico . . . *Journal of Mammalogy,* **59:** 519–530.

4.64 Hanák, V. & Horácek, I. 1984. Some comments on the taxonomy of *Myotis daubentoni* (Kuhl, 1819) *Myotis,* **21–22:** 7–19.

4.65 Dolan, P.G. & Carter, D.C. 1979. Distributional notes and records for Middle American Chiroptera. *Journal of Mammalogy,* **60:** 644–649.

4.66 Smithers, R.H.N. 1983. *The mammals of the Southern African subregion.* University of Pretoria, 736 pp.

4.67 Palmeirim, J.M. 1979. First record of *Myotis myotis* on the Azores Islands . . . *Archivos do Museu Bocage.* (2) *Notas e suplementos,* **7,** no. 46: 1–2.

4.68 Genoways, H.H. & Williams, S.L. 1979. Notes on bats . . . from Bonaire and Curaçao, Dutch West Indies. *Annals of the Carnegie Museum,* **48:** 311–321.

4.69 Kobayashi, T., Maeda, K. & Harada, M. 1980. Studies on the small mammal fauna of Sabah, East Malaysia. I. Order Chiroptera and genus *Tupaia. Contributions from the Biological Laboratory Kyoto University,* **26:** 67–82.

4.70 Yoshiyuki, M. 1984. A new species of *Myotis* . . . from Hokkaido, Japan. *Bulletin of the National Science Museum, Tokyo,* Ser.A (Zool.), **10**(3): 153–158.

4.71 Menu, H. 1984. Révision du statut de *Pipistrellus subflavus* (F. Cuvier, 1832). Proposition d'un taxon generique nouveau: *Perimyotis* nov. gen. *Mammalia,* **48:** 409–416.

4.72 Koopman, K.F. 1973. Systematics of Indo-Australian *Pipistrellus. Periodicum Biologorum,* **75:** 113–116.

4.73 Harrison, D.L. 1979. A new species of pipistrelle bat (*Pipistrellus*: Vespertilionidae) from Oman. *Mammalia,* **43:** 573–576.

4.74 Robbins, C.B. 1980. Small mammals from Togo and Benin. I. Chiroptera. *Mammalia,* **44:** 83–88.

4.75 Hanák, V. & Gaisler, J. 1983. *Nyctalus leisleri* (Kuhl, 1818), une espèce nouvelle pour le continent africain. *Mammalia,* **47:** 585–587.

4.76 Jones, G.S. 1983. Ecological and distributional notes on mammals from Vietnam, including the first record of *Nyctalus. Mammalia,* **47:** 339–344.

4.77 Kock, D. 1981. Zur Chiropteren-Fauna von Burundi (Mammals). *Senckenbergiana Biologica,* (1980), **61:** 329–336.

4.78 Hill, J.E. & Evans, P.G.H. (1985). A record of *Eptesicus fuscus* . . . from Dominica, West Indies. *Mammalia,* **49:** 133–136.

4.79 DeBlase, A.F. 1980. The bats of Iran: systematics, distribution, ecology. *Fieldiana: Zoology,* (N.S.), no. 4 (Pub. 1307): i–xvii, 1–424.

4.80 Kock. D. 1981. *Philetor brachypterus* auf Neu-Britannien und dem Philippinen . . . *Senckenbergiana Biologica,* (1980), **61:** 313–319.

4.81 Koopman, K.F. 1983. A significant range extension of *Philetor* . . . with remarks on geographical variation. *Journal of Mammalogy,* **64:** 525–526.

4.82 Peterson, R.F. 1982. A new species of *Glauconycteris* from the east coast of Kenya . . . *Canadian Journal of Zoology,* **60:** 2521–2525.

4.83 Koopman, K.F. 1971. Taxonomic notes on *Chalinolobus* and *Glauconycteris* . . . *American Museum Novitates,* no. 2451: 1–10.

4.84 Hill, J.E. 1974. A review of *Scotoecus* . . . *Bulletin of the British Museum* (*Natural History*), (Zool.), **27:** 167–188.

4.85 Kitchener, D.J. & Caputi, N. 1985. Systematic revision of Australian *Scoteanax* and *Scotorepens* . . . *Records of the Western Australian Museum,* **12:** 85–146.

4.86 Baker, R.J. 1984. A sympatric cryptic species of mammal: a new species of *Rhogeessa* . . . *Systematic Zoology*, **33**: 178–183.

4.87 Hill, J.E. 1980. The status of *Vespertilio borbonicus* E. Geoffroy, 1803 . . . *Zoologische Mededeelingen, Leiden*, **55**: 287–295.

4.88 Silva Taboada, G. 1979. *Los murcielagos de Cuba*. Editorial Academica: Havana, xiv + 423 pp.

4.89 Koopman, K.F. 1984. Taxonomic and distributional notes on tropical Australian bats. *American Museum Novitates*, no. 2778: 1–48.

4.90 Peterson, R.L. 1981. Systematic variation in the *tristis* group of the bent-winged bats of the genus *Miniopterus*. *Canadian Journal of Zoology*, **59**: 828–843.

4.91 Maeda, K. 1982. Studies on the classification of *Miniopterus* in Eurasia, Australia and Melanesia. *Honyurui Kagaku* (*Mammalian Science*), Suppl. no. 1: 1–176.

4.92 Peterson, R.L. 1981. The systematic status of *Miniopterus australis* and related forms. *Bat Research News*, **22**: 48.

4.93 Melville, D.S. 1983. Notes on small mammals from Chiang Mai Province, including two species new to Thailand. *Natural History Bulletin of the Siam Society*, **31**: 157–162.

4.94 Yoshiyuki, M. 1983. A new species of *Murina* from Japan . . . *Bulletin of the National Science Museum. Tokyo*, Ser.A (Zool.), **9**: 141–148.

4.95 Engstrom, M.D. & Wilson, D.E. 1981. Systematics of *Antrozous dubiaquercus* . . . with comments on the status of *Bauerus* Van Gelder. *Annals of the Carnegie Museum*, **50**: 371–383.

4.96 Hill, J.E. & Koopman, K.F. 1981. The status of *Lamingtona lophorhina*. *Bulletin of the British Museum* (*Natural History*), (Zool.), **41**: 275–278.

4.97 Hill, J.E. & Pratt, 1981. A record of *Nyctophilus timoriensis* . . . from New Guinea. *Mammalia*, **45**: 264–266.

4.98 Daniel, M.J. & Williams, G.R. 1984. A survey of the distribution, seasonal activity and roost sites of New Zealand bats. *New Zealand Journal of Ecology*, **7**: 9–25.

4.99 Hill, J.E. & Daniel, M.J. (1985). Systematics of the New Zealand short-tailed bat *Mystacina*. *Bulletin of the British Museum* (*Natural History*), (Zool.) **48**: 279–300.

4.100 Freeman, P.W. 1981. A multivariate study of the family Molossidae . . . *Fieldiana, Zoology*, N.S. no. 7 (Pub. 1316): i–vii, 1–173.

4.101 Legendre, S. 1984. Étude odontologique des réprésentants actuels du groupe *Tadarida* . . . *Revue Suisse de Zoologie*, **91**: 399–442.

4.102 El-Rayah, M.A. 1981. A new bat of the genus *Tadarida* . . . from West Africa. *Life Sciences Occasional Papers, Royal Ontario Museum*, no. 36: 1–10.

4.103 Ibáñez, C. 1980. Descripcion de un nuevo género de quiroptero Neotropical de la familia Molossidae. *Doñana Acta Vertebrata*, **7**: 104–111.

4.104 Williams, S.L. & Genoways, H.H. 1980. Results of the Alcoa Foundation – Suriname Expeditions. IV. A new species of bat of the genus *Molossops* . . . *Annals of the Carnegie Museum*, **49**: 487–498.

4.105 Carter, D.C. & Dolan, P.G. 1978. Catalogue of type specimens of Neotropical bats in selected European Museums. *Special Publications. The Museum, Texas Tech University*, no. 15: 1–136.

4.106 Williams, S.L. & Genoways, H.H. 1980. Results of the Alcoa Foundation – Suriname Expeditions. II. Additional records of bats from Suriname. *Annals of the Carnegie Museum*, **49**: 213–236.

4.107 Zyll de Jong, C.G. van 1984. Taxonomic relationships of Nearctic small-footed bats of the *Myotis leibii* group . . . *Canadian Journal of Zoology*, **62**: 2519–2526.

4.108 Robbins, C.B., De Vree, F. & Cakenberghe, V. van 1984. A review of the systematics of the African bat genus *Scotophilus* . . . *Annales Musée Royal de l'Afrique Centrale*, (Sci. Zool.), 237: 25.

4.109 Robbins, C. 1983. A new high forest species in the African bat genus *Scotophilus* . . . *Annales Musée Royal de l'Afrique Centrale*, (Sci. Zool.), 237: 19–24.

4.110 Bergmans, W. & Rozendaal, F.G. 1988. Notes on a collection of bats from Sulawesi and some off-lying islands (Mammalia, Megachiroptera). *Zoologische Verhandelingen*, Leiden, no. 248: 1–74.

4.111 Bergmans, W. 1988, 1989. Taxonomy and biogeography of African fruit bats (Mammalia: Microchiroptera). 1. General introduction; material and methods;

results; the genus *Epomophorus* Bennett, 1836. 2. The genera *Micropteropus* Matschie, 1899 ... *Beaufortia*, **38**: 75–146; **39**: 89–153.

4.112 Payne, J. *et al.* 1985. *A field guide to the mammals of Borneo.* Kota Kinabalu: Sabah Society/World Wildlife Fund Sabah.

4.113 Boeadi & Bergmans, W. 1987. First record of *Dobsonia minor* (Dobson, 1879) from Sulawesi, Indonesia (Mammalia: Megachiroptera). *Bulletin, Zoologisch Museum, Universiteit van Amsterdam*, **11**: 69–74.

4.114 Cakenberghe, V. van & De Vree, F. 1985. Systematics of African *Nycteris* (Mammalia: Chiroptera). *Proceedings of the International Symposium on African Vertebrates*, Bonn, 53–80.

4.115 Hill, J.E. & Schlitter, D.A. 1982. A record of *Rhinolophus arcuatus* (Chiroptera: Rhinolophidae) from New Guinea, with the description of a new subspecies. *Annals of the Carnegie Museum*, **51**: 455–464.

4.116 Hill, J.E. 1986. A note on *Rhinolophus pearsonii* Horsfield, 1851 and *Rhinolophus yunanensis* Dobson, 1872 (Chiroptera: Rhinolophidae). *Journal of the Bombay Natural History Society*, **83**, (Suppl.): 12–18.

4.117 Hill, J.E. 1985. Records of bats (Chiroptera) from New Guinea, with the description of a new species of *Hipposideros* (Hipposideridae). *Mammalia*, **49**: 525–535.

4.118 Hill, J.E. *et al.* 1986. The taxonomy of leaf-nosed bats of the *Hipposideros bicolor* group (Chiroptera: Hipposideridae) from southeastern Asia. *Mammalia*, **50**, 536–540.

4.119 Webster, D.W. & Handley, C.O. 1986. Systematics of Miller's long-tongued bat, *Glossophaga longirostris*, with description of two new subspecies. *Occasional Papers. The Museum, Texas Tech University*, no. 100: 1–22.

4.120 Gardner, A.L. 1986. The taxonomic status of *Glossophaga morenoi* Martinez & Villa, 1938 (Mammalia: Chiroptera: Phyllostomidae). *Proceedings of the Biological Society of Washington*, **99**: 489–492.

4.121 Hill, J.E. 1985. The status of *Lichonycteris degener* Miller, 1931 (Chiroptera: Phyllostomidae). *Mammalia*, **49**: 579–582.

4.122 Handley, C.O. 1987. New species of mammals from northern South America: fruit-eating bats *Artibeus* Leach. *Fieldiana: Zoology*, **39**: 163–172.

4.123 Hill, J.E. & Harrison, D.L. 1987. The baculum in the Vespertilioninae (Chiroptera: Vespertilionidae) with a systematic review, a synopsis of *Pipistrellus* and *Eptesicus*, and the descriptions of a new genus and subgenus. *Bulletin of the British Museum (Natural History), (Zool.)*, **52**: 225–305.

4.124 Menu, H. 1987. Morphotypes dentaires actuels et fossiles des Chiroptères vespertilionines. 2eme partie: implications systematiques et phylogeniques. *Palaeovertebrata*, **17**, (3): 77–150.

4.125 Schlitter, D.A. & Aggundey, I.R. 1986. Systematics of African bats of the genus *Eptesicus* (Mammalia: Vespertilionidae). 1. Taxonomic status of the large serotines of eastern and southern Africa. *Cimbebasia*, Ser. A, **8**: 167–174.

4.126 Blood, B.R. & McFarlane, D.A. 1988. Notes on some bats from northern Thailand, with comments on the subgeneric status of *Myotis altarium*. *Zeitschrift für Säugetierkunde*, **53**, 276–280.

4.127 Strelkov, P.P. 1983. *Myotis mystacinus* and *Myotis brandti* in the USSR and interrelations of these species. Part. 2. *Zoologicheskii Zhurnal*, **62**, (2): 259–270.

4.128 Kitchener, D.J. *et al.* 1986. Revision of Australo-Papuan *Pipistrellus* and *Falsistrellus* (Microchiroptera: Vespertilionidae). *Records of the Western Australian Museum*, **12**: 435–495.

4.129 Kitchener, D.J. *et al.* 1987. Revision of Australian *Eptesicus* (Microchiroptera: Vespertilionidae). *Records of the Western Australian Museum*, **13**, 427–500.

4.130 Makin, D. & Harrison, D.L. 1988. Occurrence of *Pipistrellus ariel* Thomas, 1894 (Chiroptera: Vespertilionidae) in Israel. *Mammalia*, **52**: 419–422.

4.131 Francis, C.M. & Hill, J.E. 1986. A review of the Bornean *Pipistrellus* (Mammalia: Chiroptera). *Mammalia*, **50**: 43–55.

4.132 Thorn, E. 1988. The status of *Pipistrellus capensis notius* (= *Scabrifer notius* G. M. Allen, 1908) as prior synonym for *Eptesicus melckorum* A. Roberts, 1919. *Mammalia*, **52**: 371–377.

4.133 Happold, D.C.D. *et al.* 1987. The bats of Malawi. *Mammalia*, **51**: 337–414.

4.134 Ibáñeza, C. & Valverde, J.A. 1985. Taxonomic status of *Eptesicus platyops* (Thomas, 1901) (Chiroptera: Vespertilionidae). *Zeitschrift für Säugetierkunde*, **50**: 241–242.

4.135 Strelkov, P.P. 1986. The Gobi bat (*Eptesicus gobiensis* Bobrinskoy, 1926), a new species of chiropterans of Palaearctic fauna. *Zoologicheskii Zhurnal*, **65**: 1103–1108.

4.136 Heaney, L.R. & Alcala, A.C. 1986. Flat-headed bats (Mammalia: *Tylonycteris*) from the Philippine Islands. *Silliman Journal*, **33**: 117–123.

4.137 Ibáñez, C. & Fernández, R. 1985. Systematic status of the long-eared bat *Plecotus teneriffae* Barrett-Hamilton, 1907 (Chiroptera: Vespertilionidae). *Säugetierkundliche Mitteilungen*, **32**: 143–149.

4.138 Koopman, K.F. 1984. A progress report on the systematics of African *Scotophilus* (Vespertilionidae). *Proceedings of the Sixth International Bat Research Conference* (Eds. Okon, E.E. & A.E. Caxton-Martin, University of Ife Press, Ife-Ife, Nigeria), 102–113.

4.139 Robbins, C.B. *et al.* 1985. a systematic revision of the bat genus *Scotophilus* (Vespertilionidae). *Annales Musee Royal de l'Afrique Centrale*, (Sci. Zool.), **2467**: 53–84.

4.140 Martin, C.O. & Schmidly, D.J. 1982. Taxonomic review of the pallid bat, *Antrozous pallidus* (Le Conte). *Special Publications. The Museum, Texas Tech University*, no. 18: 1–48.

4.141 Hermanson, J.W. & O'Shea, T.J. 1983. *Antrozous pallidus*. *Mammalian Species*, no. 213: 1–8.

4.142 Maeda, K. 1980. Review on the classification of little tube-nosed bats, *Murina aurata* group. *Mammalia*, **44**, 531–551.

4.143 Hill, J. E. & Rozendaal, F.G. 1989. Records of bats (Microchiroptera) from Wallacea. *Zoologische Verhandelingen*, Leiden, **63**: 97–122.

4.144 Koopman, K.F. 1989. Systematic notes on Liberian bats. *American Museum Novitates*, no. 2946: 1–11.

4.145 Varty, N. & Hill, J.E. 1988. Notes on a collection of bats from the riverine forests of the Jubba Valley, southern Somalia, including two species new to Somalia. *Mammalia*, **52**: 533–540.

4.146 Hill, J.E. & Zubaid, A. 1989. The Dyak leaf-nosed bat, *Hipposideros dyacorum* Thomas, 1902 (Chiroptera: Hipposideridae) in Peninsular Malaysia. *Mammalia*, **53**: 307–308.

4.147 Hill, J.E. & Topál, G. (in press). Records of Marshall's horseshoe bat, *Rhinolophus marshalli* Thonglongya, 1973 (Chiroptera: Rhinolophidae) from Vietnam. *Mammalia*.

4.148 Baker, R.J., Dunn, C.G. & Nelson, K. 1988. Allozymic study of the relationships of *Phylloderma* and four species of *Phyllostomus*. *Occasional Papers. The Museum, Texas Tech University*, no. 125: 1–14.

4.149 Arita, H.T. & Humphrey, S.R. 1988. Revision taxonomica de los murcielagos magueyeros del genero *Leptonycteris* (Chiroptera: Phyllostomidae). *Acta Zoologica Mexicana* N.S. **29**, 1–60.

4.150 Koopman, K.F. & Danforth, T.N. 1989. A record of the tube-nosed bat (*Murina florium*) from western New Guinea (Irian Jaya) with notes on related species (Chiroptera, Vespertilionidae). *American Museum Novitates*, no. 2934: 1–5.

4.151 Hill, J.E., Harrison, D.L. & Jones, T.S. 1988. New records of bats (Microchiroptera) from Nigeria. *Mammalia*, **52**: 590–592.

4.152 Parnaby, H.E. 1987. Distribution and taxonomy of the Long-eared bats, *Nyctophilus gouldi* Tomes, 1858 and *Nyctophilus bifax* Thomas, 1915 (Chiroptera: Vespertilionidae) in eastern Australia. *Proceedings Linnaean Society New South Wales* (1986), **109**: 153–174.

5. Primates

5.1   Napier, P.H. 1976, 1981, 1985. *Catalogue of primates in the British Museum (Natural History)... Part 1: families Callitrichidae and Cebidae: Part 2: family Cercopithecidae, subfamily Cercopithecinae: Part 3: family Cercopithecidae, subfamily Colobinae.* London: British Museum (Natural History).

5.2   Napier, J.R. & Napier, P.H. 1985. *Natural history of the primates.* London: British Museum (Natural History), 200 pp.

5.3   Wolfheim, J.H. 1983. *Primates of the world: distribution, abundance and conservation.* Chur (Switzerland): Harwood, 831 pp.

5.4   Tattersall, I. 1982. *The primates of Madagascar.* New York: Columbia University Press, xiv + 382 pp.

5.5   Rosenberger, A.L. & Coimbra-Filho, A.F. 1984. Morphology, taxonomic status and affinities of the lion tamarins, *Leontopithecus . . . Folia Primatologica,* **42**: 149–179.

5.6   Hershkovitz, P. 1983. Two new species of night monkeys, genus *Aotus* (Cebidae, Platyrrhini): a preliminary report on *Aotus* taxonomy. *American Journal of Primatology,* **4**: 209–243.

5.7   Hershkovitz, P. 1984. Taxonomy of squirrel monkeys genus *Saimiri . . . American Journal of Primatology,* **7**: 155–210.

5.8   Thys van den Audenaerde, D.F.E. 1977. Description of a monkey-skin from East-Central Zaire as a probably new monkey-species . . . *Revue de Zoologie Africaine,* **91**: 1000–1010.

5.9   Brandon-Jones, D. 1984. Colobus and leaf monkeys. Pp. 398–408 in Macdonald, D. (Ed.) *The encyclopaedia of mammals,* vol. 1. London: Allen & Unwin.

5.10 Meier, B. *et al.* 1987. A new species of *Hapalemur* . . . from South East Madagascar. *Folia Primatologia,* **48**: 211–215.

5.11 Musser, G.G. & Dagosto, M. 1987. The identity of *Tarsius pumilus. American Museum Novitates,* **2867**: 53 pp.

5.12 Thorington, R.W. 1988. Taxonomic status of *Saguinus tripartitus. American Journal of Primatology,* **15**: 367–371.

5.13 Hershkovitz, P. 1988. Origin, speciation and distribution of South American titi monkeys, genus *Callicebus . . . Proceedings of the Academy of Natural Science, Philadelphia,* **140**: 240–272.

5.14 Ayres, J.M. 1985. On a new species of squirrel monkey, genus *Saimiri . . . Papeis Avulsos de Zoologia, S Paulo,* **36**: (14): 147–164.

5.15 Hershkovitz, P. 1987. The taxonomy of South American sakis, genus *Pithecia . . . American Journal of Primatology,* **12**: 387–468.

5.16 Harrison, M.J.S. 1988. A new species of guenon (genus *Cercopithecus*) from Gabon. *Journal of Zoology, London,* **215**: 561–575.

5.17 Weitzel, V. & Groves, C.P. 1985. The nomenclature and taxonomy of the colobine monkeys of Java. *International Journal of Primatology,* **6**: 399–409.

5.18 Marshall, J. & Sugardjito, J. 1986. Gibbon systematics, pp. 137–185 in *Comparative primate biology, vol. 1: Systematics, evolution and anatomy.* Liss.

5.19 Groves, C.P. & Eaglen, R.H. 1988. Systematics of the Lemuridae . . . *Journal of Human Evolution,* **17**: 513–438.

5.20 Simons, E.L. 1988. A new species of *Propithecus* (Primates) from northeast Madagascar. *Folia Primatologia* **50**: 143–151.

6. Carnivora

6.1   Crawford-Cabral, J. 1982. The classification of the genets . . . *Boletim da Sociedade Portuguesa de Ciencias Naturais,* **20**: 97–114.

6.2   Goldman, C.A. 1984. Systematic revision of the African mongoose genus *Crossarchus* . . . *Canadian Journal of Zoology,* **62**: 1618–1630.

6.3   Gao, Y. *et al.* 1987. *Fauna Sinica/Mammalia/Vol. 8: Carnivora.* Beijing: Science Press, 377 pp.

6.4   Mayr, E. 1986. Uncertainty in science: is the giant panda a bear or a racoon? *Nature, London,* **323**: 769–771.

6.5   Wozencraft, W.C. 1986. A new species of striped mongoose from Madagascar. *Journal of Mammalogy,* **67**: 561–571.

6.6   Watson, J.P. & Dippenaar, N.J. 1987. The species limits of *Galerella sanguinea . . ., G. pulverulenta* and *G. nigrita. Navorsinge van den Nasionale Museum, Bloemfontein,* **5**: 356–414.

7. Pinnipedia, Cetacea, Sirenia

7.1   Ridgeway, S.H. & Harrison, R. (eds) 1981–. *Handbook of marine mammals;* vol. 1 (1981) *The walrus, sea lions, fur seals and sea otters,* 235 pp.; vol. 2 (1982) *Seals,* 359 pp.; vol. 3 (1985). *The sirenians and baleeen whales,* 362 pp. London etc.: Academic Press.

7.2   King, J.E. 1983. *Seals of the world,* 2nd ed. London; Oxford: British Museum (Natural History), Oxford University Press, 240 pp.

7.3   Wyss, A.R. 1988. Evidence from flipper structure for a single origin of pinnipeds. *Nature, London,* **334**: 427–428.

7.4   Baker, A.N. 1983. *Whales and dolphins of New Zealand and Australia: an identification guide.* Wellington: Victoria University Press, 133 pp.

7.5   Watson, L. 1981. *Sea guide to whales of the world.* London: Hutchinson, 302 pp.

7.6   Perrin, W.F. *et al.* 1987. Revision of the spotted dolphins, *Stenella* spp. *Marine Mammal Science,* **3**: 99–170.

7.7   Berzin, A.A. & Vladimirov, V.L. 1983. A new species of killer whale (Cetacea, Delphinidae) from the Antarctic waters. *Zoologicheskii Zhurnal,* **62**: 287–295.

8. Ungulates, Pholidota

8.1   Wetzel, R.M. *et al.* 1975. *Catagonus,* an 'extinct' peccary, alive in Paraguay. *Science,* **189**: 379–381.

8.2   Li, Z.-x. 1981. On a new species of musk deer from China. *Zoological Research,* **2**: 157–160 (Chinese), 161 (English).

8.3   Groves, C.P. & Grubb, P. 1982. The species of muntjak (genus *Muntiacus*) in Borneo . . . *Zoologische Mededeelingen. Leiden,* **56**: 203–214.

8.4   Grubb, P. & Groves, C.P. 1983. Notes on the taxonomy of the deer . . . of the Philippines. *Zoologischer Anzeiger,* **210**: 119–144.

8.5   Yalden, D.W. 1978. A revision of the dik-diks of the subgenus *Madoqua* (*Madoqua*). *Monitore Zoologico Italiano,* N.S. Suppl. **11**: 245–264.

8.6   Patterson, B. 1979. Pholidota and Tubulidentata. *In* Maglio, V.J. & Cooke, H.B.S. (Eds.) *Evolution of African mammals.* Harvard.

8.7   Groves, C.P. & Lay, D.M. 1985. A new species of the genus *Gazella* . . . from the Arabian Peninsula. *Mammalia,* **49**: 27–36.

8.8   Schlawe, L. 1986. Seltene Pfleglinge aus Dschungarei und Mongolei: Kulane, *Equus hemionus hemionus* Pallas, 1775. *Zoologische Garten,* **56**, 299–323.

8.9   Sokolov, V.E. & Gromov, V.S. 1986. A new taxonomy of modern roe deer (*Capreolus* Gray, 1821). *Izvestiya Akademii Nauk SSSR, Ser. Biol.,* **4**: 485–493.

8.10   Van Zyll de Yong, C.G. 1986. A systematic study of recent bison. *Publications in Natural Sciences, National Museum of Natural Science, Canada,* no. 6: viii + 69 pp.

8.11   Furley, C.W. *et al.* 1988. Systematics and chromosomes of the Indian gazelle, *Gazella bennetti* (Sykes, 1831). *Zeitschrift für Säugetierkunde,* **53**: 48–54.

8.12   Masini, F. & Lovari, S. 1988. Systematics, phylogenetic relationships and dispersal of the chamois (*Rupicapra* spp.). *Quaternary Research,* **30**: 339–349.

9. Rodentia

9.1   Cao Van Sung 1984. Inventaire des rongeurs du Vietnam. *Mammalia,* **48**: 391–395.

9.2   Watts, C.H.S. & Aslin, H.J. 1981. *The rodents of Australia.* London etc: Angus & Robertson, 321 pp.

9.3   Heaney, L.R. 1985. Systematics of oriental pygmy squirrels of the genera *Exilisciurus* and *Nannosciurus* . . . *Miscellaneous Publications of the Museum of Zoology, University of Michigan,* **170**: 58 pp.

9.4   Menzies, J.I. & Dennis, E. 1979. *Handbook of New Guinea rodents.* Wau, New Guinea: Wau Ecology Institute, 68 pp.

9.5   Patterson, B.D. 1984. Geographic variation and taxonomy of Colorado and Hopi chipmunks (genus *Eutamias*). *Journal of Mammalogy,* **65**: 442–456.

9.6   Levenson, H. & Hoffmann, R.S. 1984. Systematic relationships among taxa in the Townsend chipmunk group. *Southwestern Naturalist,* **29**: 157–168.

9.7   Saha, S.S. 1981. A new genus and a new species of flying squirrel (Mammalia: Rodentia: Sciuridae) from northwestern India. *Bulletin of the Zoological Survey of India,* **4**(3): 331–336.

9.8   Chakraborty, S. 1981 Studies on *Sciuropterus baberi* Blyth . . . *Proceedings of the Zoological Society.* Calcutta, **32**: 57–63.

9.9   Rogers, D.S. & Schmidly, D.J. 1982. Systematics of spiny pocket mice (genus *Heteromys*) . . . *Journal of Mammalogy,* **63**: 375–386.

9.10 Locks, M. 1981. Nova especie de *Oecomys* de Brasilia, DF, Brasil (Cricetidae, Rodentia). *Boletim do Museu Nacional do Rio de Janeiro, NS,* **300**: 7pp.

9.11 Barbour, D.B. & Humphrey, S.R. 1982. Status of the silver rice rat (*Oryzomys argentatus*). *Florida Scientist,* **45**: 112–116.

9.12 Hutterer, R. & Hirsch, U. 1979. Ein neuer *Nesoryzomys* von der Insel Fernandina, Galapagos . . . *Bonner Zoologische Beitrage,* **30**: 276–283.

9.13 Lee, M.R. & Schmidly, D.J. 1977. A new species of *Peromyscus* (Rodentia: Muridae) from Coahuila, Mexico. *Journal of Mammalogy,* **58**: 263–268.

9.14 Patterson, B.D. Gallardo, M.H. & Freas, K.E. 1984. Systematics of mice of the subgenus *Akodon* . . . in southern South America . . . *Fieldiana Zoology, NS,* **23**: 16 pp.

9.15 Pine, R.H. 1976. A new species of *Akodon* from Isla de los Estados, Argentina. *Mammalia,* **40**: 63–68.

9.16 De Santis, L.J. & Justo, E.R. 1980. *Akodon (Abrothrix) mansoensis* sp. nov., un nuevo 'raton lanoso' de la provincia de Rio Negro, Argentina . . . *Neotropica,* **26**: 121–127.

9.17 Massoia, E. 1979. Descripcion de un genero y especie nuevos, *Bibimys torresi . . . Physis Buenos Aires,* **38c** (95): 1–7.

9.18 Pearson, O.P. 1984. Taxonomy and natural history of some fossorial rodents of Patagonia, southern Argentina. *Journal of Zoology,* London, **202**: 225–237.

9.19 Olds, N. & Anderson, S. 1987. Notes on Bolivian mammals 2. Taxonomy and distribution of rice rats of the subgenus *Oligoryzomys. Fieldiana Zoology,* **39**: 261–281.

9.20 Kovalskaya, Yu. & Sokolov, V.E. 1980. *Microtus evoronensis* sp. n. . . . from the lower Amur territory. *Zoologicheskii Zhurnal,* **59**: 1409–1416.

9.21 Malygin, V.M. 1983. [*Systematics of the common voles*]. Moscow: Izdat. Nauka, 208 pp.

9.22 Lay, D.M. 1983. Taxonomy of the genus *Gerbillus . . . Zeitschrift fur Säugetierkunde,* **48**: 329–354.

9.23 Dieterlen, F. & Rupp, H. 1978. *Megadendromus nikolausi,* gen. nov., sp. nov., ein neuer Nager aus Athiopien. *Zeitschrift fur Säugetierkunde,* **43**: 129–143.

9.24 Hubert, B. 1978. Revision of the genus *Saccostomus . . . Bulletin of Carnegie Museum of Natural History,* **6**: 48–52.

9.25 Wang, Y., Hu, J. & Chen, K. 1980. A new species of Murinae – *Vernaya foramena* sp. nov. *Acta Zoologica Sinica,* **26**: 393–397.

9.26 Van der Straeten, E. 1975. *Lemniscomys bellieri,* a new species of Muridae for the Ivory Coast. *Revue Zoologique Africaine,* **89**: 906–908.

9.27 Van der Straeten, E. 1980. A new species of *Lemniscomys* (Muridae) from Zambia. *Annals of the Cape Provincial Museums,* **13**: 55–62.

9.28 Mishra, A.C. & Dhanda, V. 1975. Review of the genus *Millardia . . . Journal of Mammalogy,* **56**: 76–80.

9.29 Abe, H. 1983. Variation and taxonomy of *Niviventer fulvescens* and notes on *Niviventer* group of rats in Thailand. *Journal of the Mammalogical Society of Japan,* **9**: 151–161.

9.30 Musser, G.G., Marshall, J.T. & Boeadi. 1979. Definition and contents of the Sundaic genus *Maxomys* (Rodentia, Muridae). *Journal of Mammalogy,* **60**: 592–606.

9.31 Musser, G.G. 1981. Results of the Archbold Expeditions. No. 105. Notes on the systematics of Indo-Malayan murid rodents, and descriptions of new genera and species from Ceylon, Sulawezi and the Philippines. *Bulletin of the American Museum of Natural History,* **68**: 225–334.

9.32 Musser, G.G. 1981. The giant rat of Flores and its relatives east of Borneo and Bali. *Bulletin of the American Museum of Natural History,* **169**: 71–175.

9.33 Taylor, J.M. Calaby, J.H. & Van Deusen, H.M. 1982. A revision of the genus *Rattus* (Rodentia, Muridae) in the New Guinea Region. *Bulletin of the American Museum of Natural History,* **173**: 177–336.

9.34 Musser, G.G. & Newcomb, C. 1983. Malaysian murids and the giant rat of Sumatra. *Bulletin of the American Museum of Natural History,* **174**: 327–598.

9.35 Musser, G.G. 1982. Results of the Archbold Expeditions. No. 107. A new genus of arboreal rat from Luzon Island in the Philippines. *American Museum Novitates,* **2730**: 23 pp.

9.36 Musser, G.G. 1981. A genus of arboreal rat from west Java, Indonesia. *Zoologische Verhandelingen. Leiden,* **189**: 35 pp, 4 pls.

9.37 Musser G.G. 1988. Sundaic *Rattus*: definitions of *Rattus baluensis* and *Rattus korinchi*. *American Museum Novitates*, **1862**: 24 pp.

9.38 Van der Straeten, E. & Verheyen, W.N. 1978. Taxonomic notes on the West-African *Myomys* with the description of *Myomys derooi*. *Zeitschrift fur Säugetierkunde*, **43**: 31–41.

9.39 Capanna, E., Civitelli, M.V. & Ceraso, A. 1982. Karyotypes of Somalian rodent populations. 3. *Mastomys huberti*. *Monitore Zoologico Italiano, Supplement*. **16**: 141–152.

9.40 Van der Straeten, E. & Verheyen, W.N. 1981. Étude biométrique du genre *Praomys* en Cote d'Ivoire. *Bonner Zoologische Beitrage*. **32**: 249–264.

9.41 Van der Straeten, E. & Dieterlen, F. 1983. Description de *Praomys ruppi* une nouvelle espece de Muridae d'Ethiopie. *Annales. Musee Royal de l'Afrique Centrale*, Sc. Zool. **237**: 121–127.

9.42 Van der Straeten, E. 1984. Étude biométrique des genres *Dephomys* et *Stochomys* avec quelques notes taxonomiques . . . *Revue Zoologique Africaine*, **98**: 771–798.

9.43 Watts, C.H. 1976. *Leggadina lakedownensis*, a new species of murid rodent from North Queensland. *Transactions of the Royal Society of South Australia*, **100**: 105–108.

9.44 Kitchener, D.J. 1980. A new species of *Pseudomys* . . . from Western Australia. *Record of the Western Australian Museum*, **8**: 405–414.

9.45 Fox, B.J. & Briscoe, D. 1980. *Pseudomys pilligaensis*: a new species of murid rodent from the Pilliga Scrub, northern New South Wales. *Australian Mammalogy*, **3**: 109–126.

9.46 Winter, J.W. 1984. The Thornton Peak melomys, *Melomys hadrourus*, new species (Rodentia: Muridae) a new rainforest species from northeastern Queensland, Australia. *Memoirs of the Queensland Museum*, **21**: 519–540.

9.47 Vermeiren, L.J.P. & Verheyen, W.N. 1980. Notes sur les *Leggada* de Lamto, Cote d'Ivoire, avec la description de *Leggada baoulei* sp. n. . . . *Revue Zoologique Africaine*, **94**: 570–590.

9.48 Marshall, J.T. & Sage, R.D. 1981. Taxonomy of the house mouse. *Symposia of the Zoological Society of London*, **47**: 15–25.

9.49 Marshall, J. 1977. A synopsis of the Asian species of *Mus*. *Bulletin of the American Museum of Natural History*, **158**: 173–220.

9.50 Marshall, J.I. 1986. Systematics of the genus *Mus*. *Current Topics in Microbiology and Immunology*, **127**: 12–18.

9.51 Musser, G.G. 1982. Results of the Archbold Expeditions. No. 110. *Crunomys* and the small-bodied shrew rats native to the Philippine Islands and Sulawezi (Celebes). *Bulletin of the American Museum of Natural History*, **174** (1): 1–95.

9.52 Spitzenberger, F. 1983. Die Stachelmaus von Kleinasien, *Acomys cilicicus* n. sp. *Annalen des Naturhistorischen Museums. Wien*, **81**: 443–446.

9.53 Harrison, D.L. 1980. The mammals obtained in Dhofar by the 1977 Oman Flora and Fauna Society. *In* The Scientific results of the Oman Flora and Fauna Survey 1977 (Dhofar). *Journal of Oman Studies*, Special Report 2: 387–397.

9.54 Petter, F. & Roche, J. 1981. Remarques preliminaires sur la systematique des *Acomys* (Rongeurs, Muridae). *Peracomys*, sous-genre nouveau. *Mammalia*, **45** (3): 381–383.

9.55 Khajuria, H. 1981. A bandicoot rat, *Erythronesokia bunnii*, new genus, new species . . . from Iraq. *Bulletin of the Natural History Research Center. Baghdad*, **7**: 157–164.

9.56 Dennis, E. & Menzies, J.I. 1979. A chromosomal and morphometric study of Papuan tree rats *Pogonomys* and *Chiruromys* (Rodentia, Muridae). *Journal of Zoology, London*, **189**: 315–332.

9.57 Wu, D. & Deng, X. 1984. A new species of tree mice from Yunnan, China. *Acta theriologica Sinica*, **4**: 207–21. (Chinese, English summary).

9.58 Musser, G.G. & Boeadi. 1980. A new genus of murid rodent from the Komodo Islands in Nusatenggara, Indonesia. *Journal of Mammalogy*, **61**: 395–413.

9.59 Musser, G.G. & Gordon, L.K. 1981. A new species of *Crateromys* . . . from the Philippines. *Journal of Mammalogy*, **62**: 513–525.

9.60 Musser, G.G., Gordon, L.K. & Sommer, H. 1982. Species-limits in the Philippine murid, *Chrotomys*. *Journal of Mammalogy*, **63**: 514–521.

9.61 Musser, G.G. & Piik, E. 1982. A new species of *Hydromys* from western New Guinea (Irian Jaya). *Zoologische Mededeelingen. Leiden*, **5**: 153–167.

9.62 Musser, G.G. & Freeman, P.W. 1981. A new species of *Rhynchomys* . . . from the Philippines. *Journal of Mammalogy*, **62**: 154–159.

9.63 Sokolov, V.E. & Baskevich, M.I. 1988. A new species of birch mouse – *Sicista armenica* sp.n. . . . from Lesser Caucasus. *Zoologicheskii Zhurnal*, **67**: 300–304.

9.64 Sokolov, V.E. 1981. A new species of five-toed jerboa, *Allactaga nataliae* sp. n. . . . from Mongolia. *Zoologicheskii Zhurnal*, **60**: 793–795.

9.65 Vorontsov, N.N. & Shenbrot, G.I. 1984. A systematic review of the genus *Salpingotus* (Rodentia, Dipodidae), with a description of *Salpingotus pallidus* sp. n. from Kazakhstan. *Zoologicheskii Zhurnal*, **63** (5): 731–744.

9.66 Van Weers, D.J. 1983. Specific distinction in Old World porcupines. *Zoologische Garten*, **53**: 226–232.

9.67 Ximinez, A. 1981. Notas sobre el genero *Cavia* Pallas con la descripcion de *Cavia magna* sp. n. . . . *Revista Nordestina de Biologia*, 3 (special no.), 1980: 145–179.

9.68 Varona, L.S. 1979. Subgenero y especie nuevos de *Capromys* . . . *Poeyana*, **194**: 33 pp.

9.69 Kratochvil, J. *et al.* 1978. Capromyinae (Rodentia) of Cuba. 1. *Acta Scientiarum Naturalium Academiae Scientiarum Bohemoslovacae*, **12** (11): 60 pp.

9.70 Contreras, J. & Berry, L.M. 1982. *Ctenomys bonettoi*, una nueva especie de tuco-tuco de . . . Argentina . . . *Historia Natural, Mendoza*, **2** (14): 123–124.

9.71 Contreras, J. & Berry L.M. 1982. *Ctenomys argentinus*, una nueva especie de tuco-tuco de . . . Argentina . . . *Historia Natural, Mendoza*, **2** (20): 165–173.

9.72 Travi, V.H. 1981. Nota prévia sobre nova espéce do gênero *Ctenomys* . . . *Iheringia* Ser. Zool. **60**: 123–124.

9.73 Contreras, J.R. *et al.* 1977. *Ctenomys validus*, una nueva especie de 'tunduque' de la Provincia de Mendoza . . . *Physis, Buenos Aires*, C, **36** (92): 159–162.

9.74 Gardner, A.L. & Emmons, H. 1984. Species groups in *Proechimys* (Rodentia, Echimyidae) as indicated by karyology and bullar morphology. *Journal of Mammalogy*, **65** (1): 10–25.

9.75 Petter, F. 1978. Epidémiologie de la leishmaniose en Guyane francaise, en relation avec l'existance d'une espece nouvelle de rongeurs échimyidés, *Proechimys cuvieri* sp. n. *Compte Rendu Hebdomadaire des Seances de l'Academie des Sciences. Paris*, **287** (D): 261–264.

9.76 Avila-Pires, F.D. de & Wutke, M.R.C. 1981. Taxonomia e evolucao de *Clyomys* . . . *Revista Brasileira de Biologia*, **41**: 529–534.

9.77 Williams, D.F. & Mares, M.A. 1978. A new genus and species of phyllotine rodent from northwestern Argentina. *Annals of the Carnegie Museum*, **47**: 193–221.

9.78 Orlov, V.N. & Kovalskaya, Y.M. 1978. *Microtus mujanensis* sp. n. from the Vitim River Basin. *Zoologicheskii Zhurnal*, **57**: 1224–1232.

9.79 Reig, O.A. 1987. An assessment of the systematics and evolution of the Akodontini . . . *Fieldiana Zoology*, no. 39: 349–400.

9.80 Womochel, D.R. 1978. A new species of *Allactaga* . . . from Iran. *Fieldiana, Zoology*, **72**: 65–73.

9.81 Myers, P. & Carleton, M.D. 1981. The species of *Oryzomys* (*Oligoryzomys*) in Paraguay . . . *Miscellaneous Publications. Museum of Zoology, University of Michigan*, **61**: 41 pp.

9.82 Wang, Y. 1985. A new genus and species of Gliridae – *Chaetocauda sichuanensis* gen. et sp. nov. *Acta Theriologica Sinica*, **5**: 67–73 (Chinese); 74–75 (English).

9.83 Hershkovitz, P. 1987. First South American record of Coues rice rat, *Oryzomys couesi*. *Journal of Mammalogy*, **68**: 152–154.

9.84 Musser, G.G. & Newcomb, C. 1985. Definitions of Indochinese *Rattus losea* and a new species from Vietnam. *American Museum Novitates* **2814,** 32 pp.

9.85 Musser, G.G. & Williams, M.M. 1985. Systematic studies of oryzomyine rodents . . . *American Museum Novitates* **2810**, 22 pp.

9.86 Corti, M. *et al.* 1987. Multivariate morphometrics of vesper mice (*Calomys*) . . . *Zeitschrift für Säugetierkunde*, **52**: 236–242.

9.87 Olds, N. *et al.* 1987. Notes on Bolivian mammals 3. *American Museum Novitates*, **2890**: 17 pp.

9.88 Hinojosa, P.F. *et al.* 1987. Two new species of *Oxymycterus* (Rodentia) from Peru and Bolivia. *American Museum Novitates*, **2898**: 17 pp.

9.89 Yañez, J.L. *et al.* 1987. New records and current status of *Euneomys* in southern South America. *Fieldiana Zoology*, **39**: 283–287.

9.90 Voss, R.S. 1988. Systematics and ecology of ichthyomyine rodents. *Bulletin American Museum of Natural History*, **188**: 259–493.

9.91 Nevo, E. *et al.* 1987. Allozyme differentiation and systematics of the endemic subterranean mole rats of South Africa. *Biochemical Systematics and Ecology*, **15**: 489–502.

9.92 Orlov, V.N. & Malygin, V.M. 1988. A new species of hamsters – *Cricetulus sokolovi* sp.n. . . . from People's Republic of Mongolia. *Zoologicheskii Zhurnal*, **67**: 304–308.

9.93 Ross, P.D. 1988. The taxonomic status of *Cansumys canus*. *Abstracts, Symposium of Asian Pacific Mammalogy, Beijing*.

9.94 Galkina, L.I. & Epifantseva, L.Yu. 1988. A new species of voles from the Transbaikal area. *Vestnik Zoologii*, **2**: 30–33.

9.95 Rossolima, O. L. *et al.* 1988. Variability and taxonomy of Mountain voles (*Alticola* s.str) from Mongolia, Tuva, Baical region and Altai. *Zoologicheskii Zhurnal*, **67**: 426–437.

9.96 Roguin, L. de 1988. Notes sur quelques mammifères du Baluchistan iranien. *Revue Suisse de Zoologie*, **95**: 595–606.

9.97 Yakimenko, L.V. & Lyapunova, E.A. 1986. Cytogenetic corroboration of belonging of northern mole vole from Turkmenia to *Ellobius talpinus* s.str. *Zoologicheskii Zhurnal*, **65**: 946–949.

9.98 Bates, P.J.J. 1988. Systematics and zoogeography of *Tatera* of north-east Africa and Asia. *Bonner Zoologischer Beiträge*, **39**: 265–303.

9.99 Sicard, B. *et al.* 1988. Un rongeur nouveau du Burkino Faso: *Taterillus petteri* sp. nov. *Mammalia*, **52**: 187–198.

9.100 Dyatlov, A.I. & Avanyan, L.A. 1987. Substantiation of the species rank for two subspecies of gerbils *Meriones*. *Zoologicheskii Zhurnal*, **66**: 1069–1074.

9.101 Hutterer, R. & Dieterlen, F. 1984. Two new species of the genus *Grammomys* from Ethiopia and Kenya. *Stuttgarter Beiträge zur Naturkunde*, ser. A **374**: 1–18.

9.102 Carleton, M.D. & Robbins, C.B. 1985. On the status and affinities of *Hybomys planifrons*. *Proceedings of the Biological Society of Washington*, **98**: 956–1003.

9.103 Van der Straeten, E. 1985. Note sur *Hybomys basilii* . . . *Bonner Zoologischer Beiträge*, **36**: 1–8.

9.104 Van der Straeten, E. & Hutterer, R. 1986. *Hybomys eisentrauti*, une nouvelle espèce de Muridae du Cameroun. *Mammalia*, **50**: 35–42.

9.105 Petter, F. 1986. Un ronguer nouveau du Mont Oku (Cameroun): *Lamottemys okuensis* gen. nov., sp. nov. *Cimbebasia*, A, **8**: 97–105.

9.106 Musser, G.G. & Heaney, L.R. 1985. Philippine *Rattus*: a new species from the Sulu Archipelago. *American Museum Novitates*, **2818**: 32 pp.

9.107 Van der Straeten, E. & Dieterlen, F. 1987. *Praomys misonnei*, new species of Muridae from eastern Zaire. *Stuttgarter Beiträge zur Naturkunde* ser. A, **402**: 1–11.

9.108 Robbins, C.B. & Van der Straeten, E. 1989. Comments on the systematics of *Mastomys* with the description of a new West African species. *Senckenbergiana Biologica*, **69**: 1–14.

9.109 Kitchener, D.J. 1985. Description of a new species of *Pseudomys* from Northern Territory. *Records of the Western Australian Museum*, **12**: 207–219.

9.110 Kitchener, D.J. *et al.* 1985. Redescription of *Pseudomys bolami*. *Australian Mammalogy*, **8**: 149–160.

9.111 Kitchener, D.J. & Humphreys, W.F. 1986. Description of a new species of *Pseudomys*. *Records of the Western Australian Museum*, **12**: 419–434.

9.112 Musser, G.G. *et al.* 1985. Philippine rats: a new species of *Crateromys* from Dinagat Island. *American Museum Novitates*, **2821**: 25 pp.

9.113 Sokolov, V.E. *et al.* 1986. *Sicista kasbegica* sp. n. . . . . from the basin of the Terek River. *Zoologicheskii Zhurnal*, **65**: 949–952.

9.114 Sokolov, V.E. & Shenbrot, G.I. 1987. A new species of thick-tailed three-toed jerboa, *Stylodipus sungorus* sp. n. from Western Mongolia. *Zoologicheskii Zhurnal*, **66**: 579–587.

9.115 Bohor, B.S. *et al.* 1988. First record of *Dinomys branickii* for Venezuela. *Journal of Mammalogy*, **69**: 433.

9.116 Varona, L.S. 1986. Taxones del subgenero *Mysateles* en Isla de la Juventud, Cuba. *Poeyana*, **315**, 12 pp.

9.117 Pearson, O.P. & Christie, M.I. 1985. Les tuco-tucos (genero *Ctenomys*) do los parques nacional Lanin y Nahuel Huape, Argentina. *Historia Natural*, **5**: 337–343.

9.118 Patton, J.L. & Emmons, L.H. 1985. A review of the genus *Isothrix*. *American Museum Novitates*, **2817**: 14 pp.

9.119 Emmons, L.H. 1988. Replacement name for a genus of South American rodents (Echimyidae). *Journal of Mammalogy*, **69**: 421.

9.120 Ochoa, J. *et al.* 1988. Records of bats and rodents from Venezuela. *Mammalia*, **52**: 175–180.

9.121 Schlitter, D.A. *et al.* 1985. taxonomic status of dormice (genus *Graphiurus*) from west and central Africa. *Annals of the Carnegie Museum*, **54**: 1–9.

9.122 Mezhzherin, S.V. & Zagorodnyuk, I.V. 1989. New species of mouse of the genus *Apodemus*. *Vestnik Zoologii*, **4**: 55–59.

9.123 Sokolov, V.Ye. *et al.*1989. On species status of northern birch mice *Sicista strandi* . . . *Zoologicheskii Zhurnal*, **68**: 95–106.

9.124 Kitchener, D.J. 1989. Taxonomic appraisal of *Zyzomys* (Rodentia, Muridae) with descriptions of two new species from the Northern Territory, Australia. *Record of the West Australian Museum*, **14**: 331–373.

9.125 Flannery, T. F. *et al.* 1989. Revision of the New Guinean genus *Mallomys* . . . *Record of the Australian Museum*, **41**: 83–105.

9,126 Flannery, T. F. 1988. *Pogonomys championi*, new species, a new murid . . . from montane western Papua New Guinea. *Record of the Australian Museum*, **40**: 333–342.

9.127 Storch, G. & Lütt, O. 1989. Artstatus der Alpenwüeemaus, *Apodemus alpicola* Heinrich, 1952. *Zeitschrist für Säugetierkunde*, **54**: 337–346.

9.128 Myers, P. 1989. A preliminary revision of the *varius* group of *Akodon*. Pp. 5-54 in Redford, K. H. & Eisenberg, J. F. (eds) *Mammals of the Americas: essays in honour of Ralph M. Wetzel*. Gainsville: Sandhill Crane Press.

9.129 Myers, P. & Patton, J. L. 1989. A new species of *Akodon* from the cloud forests of eastern Cochababma Department, Bolivia. *Occasional Papers of the Museum of Zoology, University of Michigan* 720:28pp.

9.130 Myers, P. & Patton, J. L. 1989. *Akodon* of Peru and Bolivia — revision of the *fumeus* group . . . *Occasional Papers of the Museum of Zoology, University of Michigan* 721:35pp.

10. Lagomorpha, Macroscelidea

10.1 Corbet, G.B. 1983. A review of classification in the family Leporidae. *Acta Zoologica Fennica*, **174**: 11–15.

10.2 Green, J.S. & Flinders, J.T. 1980. *Brachylagus idahoensis*. *Mammalian Species*, **125**: 4 pp.

10.3 Corbet, G.B. & Hanks, J. 1968. A revision of the elephant-shrews, family Macroscelididae. *Bulletin of the British Museum (Natural History)*, (Zool.), **16**: 47–111.

10.4 Wang, Y., Luo, Z. & Feng, Z. 1985. Taxonomic revision of Yunnan hare, *Lepus comus* G. Allen with description of two new subspecies. *Zoological Research*, **6**: 101–107 (Chinese), 108–109 (English).

10.5 Wang, Y-x. *et al.* 1988. A new species of *Ochotona* from Mt Gaoligong, northwest Yunnan. *Zoological Research*, **9**: 201–207.

10.6 Li, W. & Ma, Y. 1986. A new species of *Ochotona* . . . *Acta Zoologica Sinica*, **32**: 375–379.

10.7 Angermann, R. & Feiler, A. 1988. Zur Nomenklatur, Artbegrenzung und Variabilität der Hasen (Gattung *Lepus*) in westlichen Afrika . . . *Zoologische Abhandlungen, Staatliches Museum für Tierkunde Dresden*, **43**: 149–167.

Index

236   A WORLD LIST OF MAMMALIAN SPECIES

Geomys, 146
Georychus, 207
Geosciurus, 142
Geoxus, 160
Gerbillinae, 172
Gerbillurus, 174
Gerbillus, 172
Gerbils, 172
   Pampas, 162
Gerenuk, 133
Gibbons, 100
Giraffa, 129
Giraffidae, 129
Glaucomys, 145
Glauconycteris, 80
Glider, 18
Gliridae, 196
Glironia, 12
Glirurus, 197
Glis, 196
Glischropus, 77
Globicephala, 119
Glossophaga, 61
Glossophaginae, 61
Glutton, 106
Glyphonycteris, 60
Gnu, 132
Goats, 135
Golden moles, 26
Golunda, 180
Gorals, 134
Gorgon, 132
Gorilla, 101
Gracilinanus, 10
Grammogale, 105
Grammomys, 178
Grampidelphis, 119
Grampus, 119
Graomys, 161
Graphiurus, 197
Grison, 106
Grisonella, 106
Grisons, 106
Grysbok, 133
Guanaco, 126
Guemals, 129
Guenons, 98
Guereza, 99
Guinea pigs, 201
Guira, 206
Gulo, 106
Gundis, 208
Gunomys, 194
Gymnobelideus, 18
Gymnuromys, 166
Gymnurus, 27
Gyomys, 188

Habromys, 157
Hadromys, 180
Haeromys, 190
Halichoerus, 116
Hamsters, 164
Hapale, 94
Hapalemur, 92
Hapalomys, 177
Haplonycteris, 47
Hare-wallabies, 21
Hares, 209
Harpiocephalus, 84
Harpiola, 84
Harpyionycterinae, 47
Harpyionycteris, 47
Hartebeest, 132
Hedgehog-tenrec, 25
Hedgehogs, 27
Helarctos, 104
Helictis, 107
Heliophobius, 208
Heliosciurus, 139
Helogale, 111
Hemibelideus, 19
Hemicentetes, 25
Hemiechinus, 28
Hemigalus, 110
Hemionus, 123
Hemitragus, 135
Herpailurus, 113
Herpestes, 111
Herpestidae, 110
Hesperomyinae, 150
Hesperomyotis, 72
Hesperomys, 160
Hesperoptenus, 77
Heterocephalus, 208
Heterogeomys, 147
Heterohyrax, 124
Heteromyidae, 148
Heteromys, 149
Hippocamelus, 129
Hippopotamidae, 126
Hippopotamus, 126
Hipposideridae, 56
Hipposideros, 56
Hippotigris, 123
Hippotragus, 132
Histiotus, 79
Histriophoca, 116
Hodomys, 163
Hog, 125
Hog-badger, 107
Hog-deer, 128
Holochilus, 162
Hominidae, 101
Homo, 101
Honey possum, 23
Hoplomys, 206

Horses, 123
Howler, 96
Huemuls, 129
Hutias, 203
Hyaena, 112
Hyaenas, 112
Hyaenidae, 112
Hybomys, 181
Hydrochaeridae, 201
Hydrochaerus, 201
Hydrodamalis, 123
Hydromys, 196
Hydropotes, 127
Hydrurga, 116
Hyemoschus, 127
Hylobates, 100
Hylobatidae, 100
Hylochoerus, 125
Hylomys, 27
Hylomyscus, 187
Hylonycteris, 63
Hylopetes, 145
Hyomys, 179
Hyosciurus, 139
Hyperacrius, 169
Hyperoodon, 121
Hypogeomys, 166
Hypsignathus, 44
Hypsimys, 157
Hypsiprymnodon, 19
Hypsugo, 75
Hyracoidea, 124
Hyraxes, 124
Hystricidae, 199
Hystrix, 199

Ia, 79
Ibex, 135
Ichneumia, 112
Ichthyomys, 163
Ictonyx, 106
Idionycteris, 82
Idiurus, 150
Idmi, 134
Impala, 133
Indopacetus, 121
Indri, 93
Indriidae, 93
Inia, 117
Insectivora, 25
Insectivores, 25
Iomys, 146
Irenomys, 162
Isoodon, 16
Isothrix, 206
Isotus, 72
Isthmomys, 157

DATE DUE

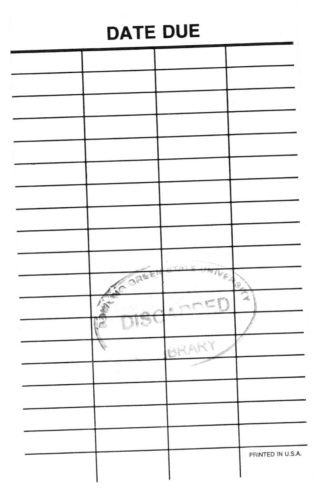